(RORTU)
Rise of Reason Through Understanding

Scientific and Philosophical Considerations

Michael Siwek

authorHOUSE®

AuthorHouse™ LLC
1663 Liberty Drive
Bloomington, IN 47403
www.authorhouse.com
Phone: 1-800-839-8640

Published by AuthorHouse   07/01/2014

ISBN: 978-1-4969-2336-3 (sc)
ISBN: 978-1-4969-2321-9 (e)

Library of Congress Control Number: 2014911925

Cover Graphics/Art Credit: Kelsey Whitney

Contents

Acknowledgements

The single most important acknowledgement for me is saved for the dedication at the end of the book. If you can, please read the book first then the dedication.

The list of acknowledgements is just too big so I'll get right to the three individuals who surprisingly tolerated a large number of hours over the last TEN years to help me write this thing. There was no one else who could withstand my persistence on talking about all this for ten years. I am still surprised with that length of time. Everyone in the office complex became painfully aware of the pure nonsense I would insist on talking about. I thank John, Enzo and Jeff for this book. We all know though that Jeff by far was the single largest contributor and in fact I know that my appropriate brain *rewiring* after surgery would not have taken place the way it did without him.

Introduction

    This book definitely needs an introduction or some suggested advice on how to read it. First it must be stated that this is a very labor intensive read in my opinion. If you read through it like a novel that will mean two things: 1. You didn't grasp the concepts and ideas as they exist in my mind, which if that is the case I did a terrible job at putting them on paper. I am quite sure that I did poor job at it but this is all I was likely capable of to date; and 2. You are likely not of these related states of human existence at the moment. All this means is that each of us has a particular collection of states which humans could hypothetically be in or have memory of being in. So for example I am very familiar with rabbit and duck hunting in my area. The experience of this is a collection of related states which even includes experiencing them to a degree of enjoyment. How much did I or do I now still enjoy experiencing these states of human existence related to rabbit and duck hunting? I also used to read science fiction fantasy books every night from the age of 10 to 28 years of age. This is a different collection of states of human existence that I am experienced in. An example of states of human existence that I have little or no experience in would be *me* making or creating art. I immensely enjoy listening (special thanks to the music <u>Rise Against</u> has provided!), looking at and marveling at essentially all art but I absolutely have a very hard time trying to do it! This is a state of human existence that I really do not 'get'. This makes me appreciate and marvel at those who can. Do you see how easy it is for me to come out and 'admit' to what I cannot or do not understand or 'get'. Having lived constantly in a state of human existence that attempts to understand how and why all of this (what this is and is not) exists, behaves and evolves is a very uncomfortable place to be in. I have been in states where simpler more specific pursuits such as sports, school and a part-time job were 'good enough'. Many people strive or hope for living a life with an easy to see and defined meaning or purpose that they or others have often chose to believe in or simply follow with no attention to belief. My problem was that I'm not sure I ever actually 'slept' at night even by the time I was 6 or 7 or was able to focus on prescribed tasks at hand without this thing always hanging in my mind of trying to understand everything. I still obviously don't understand everything and in fact I feel like I understand less than ever. Yes I have performed simpler tasks such as fishing, hunting, gardening, standard car repairs, teaching science, housework, building a house with standard wiring, plumbing etc…but none of any of these you can think of is quite the same as attempting to see and understand that which we as human systems are not *likely* ever going to be able to grasp. We can and do grasp more specific existing things. There are moments when I can almost feel my fingernails just touching it but then it is gone. I'll go back to re-immersing myself in the whole maple syrup mini operation or welding in the barn, or baking a new butter tart recipe and then talking casually at gatherings about the more specific states that us humans are *capable* of experiencing. But I am ALWAYS painfully aware of something, I'll take as a quote from one of my past favorite reads, <u>*The Life of the Bee*</u> (Maurice Maeterlinck), like "eternal truth that rests over all things like beauty in suspense". There are many people who constantly think about such things and if that is so then you will try to make sense of this book. I realized long ago that the disciplines of science (the mainstream academic pursuits) are often quite *seemingly* separate from philosophical thinking and then even philosophy is often *seemingly* separated from every day talking and speaking to each other. To reach a point where a collection of thoughts and ideas no longer have any separation yet you see how the separation lines that have been drawn is a tenuous situation at best is certainly not comfortable. It's almost like an excited kid who clearly sees a path to run to new things but realizes to do so without bringing any help along will be lost and never able to return. After retreating back to people for help, how does one clearly bring the directions and necessary ideas to light so others would even consider such 'nonsense'? This book is my attempt.

The first part that I wrote is philosophical considerations in which I purposely try to avoid including science (which was not entirely possible). I put it as the second part in this final edit. I tried to write it as nothing more than ideas to consider with no accounting in the 'real' physical world. There is going to be a noticeable (to me it was) evolution of the ideas which continues even in the scientific considerations section. I left it this way on purpose. It still makes sense to me and shows how the evolution took place. Change is always occurring and anyone who ever tried to write down philosophical types of thoughts has noticed evolution in their own thinking which never stops. But we have to write something down as long as we have it. The scientific section is worded with no effort to try and teach many concepts of science... that was a body of work I could not undertake without squashing what I most wanted to bring out in this book. I'm sorry for that and it makes the science section even more difficult to read for those who have little background in physics/physical chemistry yet there is still much that anyone could get something from. The difficulty is not an overall comprehensive read, but in **understanding** a lot of the existing scientific ideas as they apply to what is in my head.

So to sum up, how many people actually would want to read this stuff even if it was written beautifully? (Which this definitely was not) I can't imagine there would be many. To turn this around for me would be like reading a romance novel today! I would rather be beaten repeatedly with a stick! Yet millions of these books are sold, why? Because there **is** a large number of people that **are** in that collection of human states related to that type of material. That is not wrong or right; it simply is what it is at this moment in human existence.

MJS

December 31, 2013

The Existence Model

Preamble:

Information

Let's consider the word itself. Most people see that this word can be: truths, non-truths, facts, descriptions, and meanings. Others have said that information can be thought of as an answer to a question which in principle always has to be some kind of truth, non-truth, fact, description or meaning. And if this is the case then information is something that can decrease or eliminate uncertainty but then again it could increase it. Some have said that if you could 'prove' a particular piece of information was not truthful in this universe then it would just be 'nonsense'. Notice how the common use of the word information typically brings to mind the very idea or notion of what is 'real' and what is not 'real'. There always appears to be an immediate atmosphere of that which is truth and that which is not ...but we also have very little certainty about which is which. The common practice is for us to relate any piece of information we are given to what we have seen or experienced at some point in our lives ... again bringing to our mind what is true and what is not. This is *part* of how I understand information since I think there is something *more* to it. We do need to introduce a very important point to the model that is to come; I see information as something that literally exists but not simply just in spatial dimensions such as in our 'universe' in its *commonly* understood form. So for example if we 'record' some information in a book about

1

global warming you will see that in this model the <u>*physical book itself*</u> will literally have information coupled into it but to those who can read the letters they can have the intended 'meaning' (although this can occur in many different ways) delivered into and/or brought up in their mind. Don't confuse what is 'real' as far as information is concerned. The existing book will have precise and exact information coupled into its observable existence. The meanings and storage of what we call information in the pages is nothing more than what meaning we have given it to bring about various memories from our storage or bring about new understandings in our minds. Any information (what most have loosely thought of this word to be) in storage whether on pages, on disc or in our minds are *observable* representations of information. In this world that we experience we always observe the results of real information coupling with the real substance(s) of it.

Substance

Most have thought of the word substance as something that 'physically' exists. You hold up a glass or you buy one pound of sugar … and we all think of these as physical objects. In all of current science we take measurements of 'physical stuff'. Substance is the physical stuff used to make any observable outcome. More succinctly I see substance to be simply energy as far as we can tell for now. In all of physics and science that is taught the only substance we ever deal with is energy and the spatial dimensions that it can utilize. The model will allow for the possibility of other types of substances that we do not know of but could easily be the case either in our universe or in others.

Observable Outcomes

This is quite simply anything that exists in the way that we humans have always *commonly* thought of. I say commonly because most think of anything we see or interact with as something that exists. This model clearly has three types of *existence*:

1. an existence of information and forces (this will be called information/force space or I/F space)

2. an existence of substance(s) (ours is energy) and its (their) available spatial dimensions. (Ours will be called energy/spatial dimensions space or E/S space but any generic one will be a substance/spatial dimensions space or S/S space)

3. an existence or *coexistence* of when the above two couple. This will be called an observable outcome space or OO space.

Consciousness

This needs to be introduced up front so that the model will be easier to digest. The extended version of the model in this book sees consciousness as only existing in an observable outcome space and <u>CANNOT</u> exist in either I/F or an S/S space. This is crucial to make a note of here. The simple and extended model both use the idea of I/F and S/S space coupling and decoupling to yield an OO space but what consciousness is and where it resides is an essential difference between them. The extended model defines consciousness as the result of the coupling and decoupling in the yielded outcome space and hence resides only there. The simple version makes no definition of what consciousness is or where it can take place. In fact this was the reason for extending the model in the first place. A more detailed discussion of this will follow particularly in the philosophical discussions.

In the extended version of the model:

Any amount of substance that couples with I/F space and thus does and will decouple at some point in time is consciousness. What is interesting is that even a small particle such as a single electron or proton that is observable for a while, seemingly 'disappears', and then reappears like we hear about in quantum mechanics **is** a level of consciousness and it is 'processing' some quantity of information. This processing is literally it's defined existence in OO space and then it's non-existence ... this is the fundamental element of change itself. When it 'comes' back (if it does) there is always some kind of change and nothing stands 'still' nor will remain identical to what it previously was. By identical I mean more than just what the particle is or can be observed as by any other part of the whole system. Even if the particle itself appears identical, its relations to all other parts of the entire universe cannot be identical to what it was before. Any observable system that is capable of processing information either simplistically or very complex is conscious. Obviously something like a coyote compared to a single cell bacteria or single molecule has much greater capabilities for processing information and especially storing meaningful quantities of it in a representational manner so it has a much higher level of consciousness. This representational manner is how systems can 'store' their recordings/interpretations of information. For example, we humans use the bulk of our OO space brain that remains existing to store, via our representations utilizing various existing 'things', the meanings of some information that we have. What we have not been able to ascertain with 100% certainty is the validity of this information in our OO space. So for example how sure are we that gravity will always be observable according to the equations that we have? We are not 100% certain. Interestingly, this 'change' (coupling/decoupling) seems to be the necessary requirement for consciousness and hence even understanding itself.

__Comments to end the preamble__

Before outlining the model it is important to point out a few things. The first thing is that I want to try and keep it as simple as possible. I have written hundreds of pages which after thinking carefully about it I realized many of them were tangents off of numerous finer details. Not unlike a scientist finding out that they were characterizing a specific gas to such detail that they lost sight of how overall it does and can behave quite like an ideal gas. I am in no position to make detailed statements regarding any number of these ideas to come. I have no scientific proof. This model is just a group of ideas. I also in no way actually *believe* this is or must be true. It's a model and I want the starting point to remain as simple as I can leaving open the door to many considerations by those who are in a much better position to go down those paths. For example, the way consciousness was mentioned above easily could lead one to start thinking about how our entire universe could be the sum total of all conscious subsystems and how over time changes are taking place to maximize the level of consciousness that can or will exist at some point in the future....how could these subsystems be brought into the larger collective? This could lead to thinking about how this could possibly be measured or studied etc... This would be like jumping ahead to all sorts of strange real gas equations instead of sticking with understanding a simpler model first (the ideal gas one). I will purposely try to keep this simple and make some assumptions with no vigorous proof.

The Existence Model

This model clearly has three types of *existence*:

1. an existence of information and forces (this will be called information/force space or I/F space)

2. an existence of substance(s) (ours is energy) and its (their) available spatial dimensions. (Ours will be called energy/spatial dimensions space or E/S space but any generic one will be a substance/spatial dimensions space or S/S space)

3. an existence or <u>*coexistence*</u> of when the above two couple. This will be called an observable outcome space or OO space.

<u>Assumptions of The Existence Model</u>

1. I/F space couples with S/S space to yield an observable outcome in the OO space (universe) it is observed in.

Let us just assume this occurs with no statement of math describing it or debating its validity. The model will view an observable outcome space as a universe and hence invites the idea that there could be many of them. There are no constraints on how many S/S spaces there can be.

2. Observable outcomes are related with perfect precision to the 'section(s)' of I/F space that they are '*built*' from using the appropriate available substance from the S/S space that *can* be utilized.

One could view this by analogy to the house made from building materials and blueprints/minds of humans. For us the house would not likely get built exactly the way the information of ours described it before we started (especially since we do not know exactly what is true and what is not) and we simply record and reflect with notes on our blueprints any changes we've made. Outcomes always get built exactly to I/F space's *allowable* specifications …as far as what can and cannot be achieved at that particular point in time.

3. **Memory in an OO space <u>*can*</u> exist in the coupled results of I/F and S/S space but does so as representations of the interpretations of that OO space system.**

It must be noted that the concept of 'memory' inside an outcome space is important to square away what we as animals and humans definitely observe. We obviously have memories in ourselves where we store our interpretations of outcomes and use it to effect the next likely collection of outcomes for us. So ideally maybe there is the most simplistic OO space subsystem (atom? photon? …) which has no memory of its own and couples directly only with I/F space to yield outcomes. More complex systems like us which have memorized information that we have interpreted for past outcomes, can use it to perturb the next outcomes.

4. **We must envision truth from the perspective of I/F space where coupling with any S/S space to yield an outcome is truthful while the remainder which cannot couple anywhere is not truthful.**

This distinction can only be made in I/F space since the accurate total is needed. This could also be called couple-able vs non-couple-able information. Non-couple-able information is not accessible by any S/S space and thus from our current position, simply not observable in the form of actual outcomes. We then must also acknowledge that there can be information which is only couple-able with at least one S/S space (or many …) thus making it truthful in the applicable outcome space(s). We then also have to acknowledge what is thinkable versus not thinkable. Information that is thinkable versus not thinkable could be yet another classification in itself for what is couple-able and not couple-able. Simply acknowledging something as observable might not be the same as thinkable. The model will use the classification chart in Fig.-1. This requires some thinking about due to how many of us see the word 'truth'. An example would be the large group of humans that long ago did operate using information that the world was flat and one could fall off. Later, when discovered that this was not the case, it did not change the fact that systems using such information still yielded outcomes in reality that are observable. Such 'incorrect' information was then viewed as not true even though it ___did___ perturb the system as observed. In an everyday sense people view information as being true vs not true differently than what it really is; that which is couple-able to yield observable outcomes or is not couple-able. Our ___interpretation___ of these outcomes is either going to be correct or not correct to a certain level of accuracy and to some degree of precision. Obviously the interpretation of the shape of the planet was not accurate yet any system running on such an interpretation gives very predictable outcomes and these are true (since we observe the couple-able outcomes). So people who run on information that they will sail off the planet will not (voluntarily) sail out far and *this* is true as an outcome but what is not true is when people sailed out far they fell off.

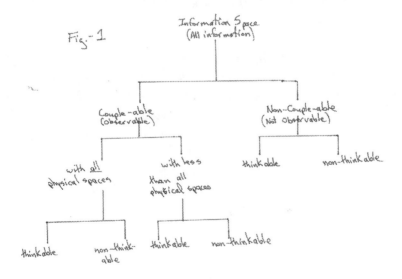

Fig. 1

Information Space
(All information)

Couple-able
(Observable)

Non-Couple-able
(Not observable)

with all physical spaces

with less than all physical spaces

thinkable

non-thinkable

thinkable

non-thinkable

thinkable

non-thinkable

5. Time and Distances only exist in an OO space and cannot be experienced in I/F or S/S space.

The coupling and decoupling of I/F and S/S spaces is the element of change itself... how nothing in an OO space can stay the same. This is the passage of time. The 'stuff' of I/F and S/S spaces does not change intrinsically or depend on time and distances. Their contents are constant. It is not possible to observe substance or information by itself in isolation. Only the coupling results are observable. This then makes our concept of space and time something that information, forces, substance and spatial dimensions themselves cannot experience. They are outside of this. This allows coupling and decoupling a great deal of freedom from such things.

The next points belong to the extended version of the Existence Model.

6. **The Extended version of the model has gaps or holes in what I/F space contains allowing for randomness or 'choice'.**

If the I/F space was completely and utterly infinite with no gaps in information then one can envision how all outcomes or any outcome space are 'known' informationally speaking. Consider a person having 10 doors out of the room they are in and they must exit. We also can with perfect perfection 'replay', exact to every detail, the same universe and situation 100 times. The simple version of the model allows for an absolute information space where for each and every trial it even has the information of the sequences of door openings. Even that information is there. The extended version of the model holds that in order to have some genuine randomness or apparent choice even information space does not have the information of each and every trial and their sequences. It could easily have information of the average outcomes such as knowing this person opens door 3 seventy seven percent of the time but not each result one by one. This would amount to a law or rule of sorts pertaining to what I/F space can contain. What is 'missing' or not allowed could be called gaps or holes.

7. **The extended version of the model states that consciousness exists only in the OO space and not in I/F or S/S space.**

After conservations with many people it often was quickly summed up by them that the model is just god and then any *genuine* discussion on their part promptly ended. True, the simple version of the model does not define consciousness nor make any statements as to where it exists. To my knowledge science has no generally *accepted* definition of consciousness nor physically described mechanism. We as humans have a large number of ideas about what consciousness is starting with something as simple as fainting or a coma then 'regaining' consciousness. Many associate active thinking and cognizance as part of consciousness. This extended version of the model defines the mechanism of consciousness, what it is and where it resides.

The Processing Model

It became important to characterize how the human mind could process information. It is very likely that there are other ways that the human mind does process information but is very difficult to observe or notice it is there and how it might work. What I have come to use as a crude model for human processing of information is the following where there are two extremes of how certain we can be when considering anything out of the Everything... 0 % certain on one end and 100% certain on the other.

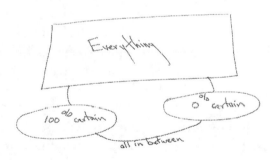

Assumptions of the Processing Model

1. **Assume we live in an existence of an infinite amount of information.**

   The key idea is any processing unit requires some kind of assumptions assumed to have a high degree of certainty or simply be true out of the infinite information so that a stable foundation for processing can occur. Once some kind of fundamental processing is set up based on assumed (or believed) truths then reasonable stability of processing can be obtained. An analogy could be the rocket ship in space where we need to provide some material out of our rocket so we can then push off of it otherwise we can't go anywhere under our control. Or another analogy of an endless open sea and sky (no sun) with no bearings at all in sight …so we let out 5 or 6 row boats in as long a line as we can so that we 'create' our own straight line bearing. In a world of infinite information most of which *might* not even be applicable to our outcome space, processing systems *have* to start off of something assumed to be true and then make alterations as we go. Some systems acknowledge they are using some information as if it was true while others treat some information as truth.

2. 'Everything' in the Processing Model is <u>all</u> of what actually informationally is.

One of the most basic assumptions that can be made is that whatever this (our world, universe, anything taking place, anything beyond etc…) is, … *is* <u>something</u>. Some philosophical debates can take the position that it is *nothing* but for this model let's assume it *is* something. The Everything in this model refers to information in its entirety since this is about processing of information. This Everything is not the actual substance itself.

3. Treat the human mind as a biological processing machine which must make decisions about how certain it is regarding the 'correctness' of any information it is aware of or else it just does what it sees as truth.

The basic idea is to think of our minds as a complex computer processor that must have some kind of fundamental basis of information to work off of (high order of certainty or truth) since we are immersed in an infinite sea of information. This Processing Model assumes that we humans have essentially one way of trying to assess, characterize, understand or interact with Everything and that is simply how certain of something we are. All human processors base further interactions off of what they currently either hold as 100% certain or have higher levels of % certainty for.

Discussion of the Processing Model ...

Let's consider a simple scenario where myself and one other human (25 years old and is an average North American person) are simply told to picture a dragon in our minds. We are not talking with each other or communicating. How certain am I that this other person pictures a creature in their mind that has many features/characteristics that are in common with my image of a dragon? Interesting...since they do appear to be an average North American 25 year old I am very confident in the 95% + range that their picture and mine have a lot in common... but I am not 100% certain. What if they were severely mentally disabled and in fact did not even have any picture for a dragon at all? I am admitting to myself that I literally do not know this but I do know that most people do have similar pictures of dragons in their minds. Another question is asked ... "How certain are you that their picture of a dragon in their mind is <u>identical</u> to yours?" I immediately start accessing a lot of other experiences and information (whether I want to or not) related to myself and other people and what has often seemed to be the case, which is that an identical image in our minds is incredibly unlikely and so I am somewhere in the 0.00001% certain that theirs is identical to mine. Note that an informational processing system seems to immediately relate all things to what it is quite certain (or is in a state of truth) about and then answers the question accordingly. So here I immediately sorted through information that I was quite certain about and then upon reviewing the question I knew I had to answer it that I was very uncertain. Just the same in principle though if the question was, "how certain are you that their picture in their mind was <u>not</u> identical to yours?" is that I would answer 99.9999% certain. The real point here is that for the same question different biological processors will give a value somewhere on this scale from 0% to 100 % certainty. It gives us an indication of how certain the processor is for a piece of information whether they approach 100% or 0%. A couple of things are important to note here. One is that a complex biological processor like us humans immediately relates a lot of other pieces of information on our hard drives to a question, task, thought, idea etc...

and ultimately come to a % certainty on it at that time. Whether you come up with 1 % certainty or 99 % certainty you obviously are quite certain about something. The closer to 50% certainty you are the closer to having no idea one way or the other you are. The other note here is what does 100% certainty inside of a processing unit mean? Does it mean that it ***genuinely*** knows something as truth out of the infinite sea of information? This gets bit interesting and I am not sure of the 'real' answer at all! What I see happening in this model is a system that does not 'think of something as true" or "thinks of this as being the truth", no, no, no, …no …They literally see it as truth. None of this thinking or pretending …. truth just is in that processor. This model makes no claim as to the validity or lack of validity of information out of the infinite sea, but simply sees that this system utilizes that information as truth and thus is not subject to questioning it so long as that system remains at 100 % certain for that piece of information.

Complex biological processors have to first have some information in their minds and then once processing of it starts it must have some means of making decisions which in turn relies on degrees of certainty it possesses. What is interesting is a calculator for instance, where wherever it encounters 1 + 1 it always give 2 unless something in it has changed. If it works the way we intended it to, it will always give the answer of 2. This processor operates where everything it ever considers is done on a basis of 100 % certainty in what it will and can do. Any processor that has even the slightest processing abilities to alter its degree of certainty in anything inside itself will be subject to change and yield differing outcomes over time. This model clearly sees the operational state of 100% certainty as being the very interesting and different thing from less than 100 % certain. In human history we have seen many examples of them in a processing sense seeing some information as 100% certain. I would call the state of having the highest level of faith or outright belief in something as being this interesting case of 100 % certainty. I've seen the debate between some who acknowledge their faith is vulnerable at times and those who clearly experience only 100 % certainty in something at all times including burning at the stake with no reservations at all. That is a very interesting state of processing information and it does not have to be only with religious topics. For instance if there was a physics conference where the latest 5 front runners for a theory of something were being discussed it becomes evident when a person is essentially running with 100 % certainty in theory number 3. When this is experienced by someone they often devote the rest of their lives committed completely to this theory number 3 and we do get many great results even if over time theory number 3 is shown to be very, very unlikely. But the point is that human systems do demonstrate very inspirational and powerful resulting outcomes derived from this informational configuration. It is a 'real' observable result that does show differences from those who only operate in a non-100% certain way. The key idea here is that a human system operating literally in a 100 % certain state about some amount of information does occur and is quite different than one which does not. The purest form of the scientific method is at no point in time to enter into 100% certainty that something is literally truth. Faith/belief based methods of human interaction with Everything simply encourages adopting some information as literal truth. Most humans clearly seem to show varying amounts of these two approaches. I've seen some people who literally use a lot of information about all things they can or cannot do from a faith/belief style of existence and thus their life and observable outcomes of what they do or do not do is quite well defined and shows little variance. And this is normal. I've also seen others who have such little certainty in doing anything that they can be quite unstable as an existing human. Then there is what I see as the much larger majority of humans which is those using significant pieces of information that they are highly certain of, but not 100 % and thus change occurs in a quasi or somewhat stable manner.

Note for the rest of the book ...

These ideas were built up over decades without even realizing it. The key starting point when I was between 7 and 9 was when I thought about how the 'truth' whether a human could move things with their mind or not must exist. The frustration was not knowing it with 100% certainty. I started to focus literally on the concept of information itself when I was approximately 35 and I even started with a variety of vague ideas seemingly completely unrelated to this one. Over the last 7 years I have become more comfortable with this idea and I must admit that many if not all the things I had thought about my whole life don't mind this model either. This does not make it real but is definitely worth considering. I will essentially have two parts to this book:

1. Scientific Considerations
2. Philosophical Considerations

Section I: Scientific Considerations

Some of the concepts or phenomena in science that are constantly used and often taken for granted in the process are space (spatial), time, inertia, energy, forces, change, distingishability and perhaps the most important: information. These are some of the fundamental concepts that I need to review first to get a feel for what is commonly thought and then attempt to identify the rabbit in the pile I know is there (the 'I know it's there' part is a leap of faith!).

Space (Spatial)

Space has particularly been viewed as some kind of existing thing or substance. Not unlike how a piece of canvas is the something that a painting can exist in or on. Growing up this made the most sense and I have found that most people on average have this type of view. Initially at least. But after really thinking about it for awhile many couldn't help realizing it is possible that the walls of a room for instance create a 'space' in the sense that we measure everything in it in relation to them. We measure or at least visually compare distances between all things. Even in physics or cosmology they must measure distances in relation to something they see otherwise what can they do? The concept of some kind of absolute reference point, object or space has gone cool in science today while we now typically see all measurements and objects themselves in relation to others. Does space actually exist in a real objective sense? Many argue that it either does or does not. Then some said it could be both and then others said neither! So what can or could we do? We start with the fact that we do not have any conclusive proof of exactly what space (spatial) is or is not. We can think up as much as we want but what is the truth for this place that we live in? Science had no choice but to come up with working definitions/ ideas for such a thing. When you look up the definition of a word like 'apple', we find quite a clear answer that does not leave much doubt (in our daily lives). We all have seen one and eaten one. But what about space? Scientists don't have any working definitions/ideas for an apple (since we don't need one …) but they certainly do for space since they have no *conclusive* idea for it. So they use an idea for the moment until something causes them to alter it. All humans naturally had a sense of up, down, left right and then back and forward which they then called three dimensional space. Math already had descriptions of such a thing and even for n-space (any number of dimensions you wish). This was used as our working definition/idea until such time that observations came to light to change it. Einstein (others already had significant pieces of the ideas but not a hammered out product of sorts) came up with the idea of space-time which now has become our working idea which is 4 dimensions (3 spatial and one time). The proof of such was not a simple and easy experiment that students can do in high school but real 'proof' did come to light. Even the satellites and your GPS require relativistic corrections to be made in order to make the results very accurate (and practically usable). The history of how science has and still does KNOWINGLY use working definitions has created quite the precedent. The fact is that we do not know what space is and what it is not. Most human systems using science today essentially are unaware of such a thing and that is normal human behavior.

Time

In my experience with people and the media, space and whether it exists in reality or not rarely gets any attention but time does. Essentially everyone I know, regardless of age has some of their own concerns and ideas about time. Everyone knows of the famous twins and a fast relativistic spaceship where one stays here and the other goes on a 50 year trip. Upon returning the travelling one only aged a few years while the other is now an old person. Then there is time travel, warp speed, teleportation in books, movies, break time banter and who knows what else. Some believe this stuff can and will be done, others say it is not possible. Say that space itself does not exist except for relations and most will look at you funny, … while mention everything will play in reverse once the universe stops expanding and contracts and no funny look at all! Does the majority know something concrete that we don't?...or do we have familiarity with common story/movie topics regarding time? What if time itself does not actually exist? I mentioned in a philosophical way that in information space time can't exist nor be experienced like we do here ...but all the information concerning time in any observable outcome space is there. This is an idea not a fact. Just like some view space as nothing more than relations between objects I can imagine viewing the arrangement, ordering or sequence of events/states of a system as getting *perceived* as time. Your perception could come from the memory of the ordering of the states and there is no such thing as going backwards in a <u>*literal sense*</u>. You then are simply 'proceeding' on a collection of states back to where you started (not as if the entire universe and all within it are literally reversing). All there could be is just *changes* in the state you are in with a memory bank of past states ... nothing more. I used to struggle with this idea but now it seems just as plausible as time actually existing like something objectively real (which is the conventional idea of time ...the majority think so). But we don't know exactly what it is and what it is not. So what do we do? We work with what we have as a working idea of time. We made calendars which counted out days in relation to our travel around the sun...365 days a year. We made our day in relation to our earth's rotation, then a relation for our hour and so on. Our memory is the only link to all the experiences we had in relation to these measures of 'time'. With no memory would you be able to think about time or relate to it in the same way we all do now? Not likely. So what is it really? We don't know. We have made up our own measure for time in seconds, minutes etc... We did the same thing for measures of distances and we found some evidence to suggest that even these relations as we have defined them are not constant. We found that time and even the distances in the same direction as your motion near the speed of light are not the same as if you stayed in the frame of reference you accelerated from. But on this planet moving at the speeds we do, the standard older view of space as just three 'constant' dimensions and 'constant' counting of seconds for time is perfectly fine. The scientific 'proof' is the reproducible practical observable results we get using these working definitions/ideas...for now.

Michael Siwek

Inertia

I just went back to quickly review what my memory has collected when inertia is *introduced* either at the grade school, high school, undergraduate, graduate or post graduate level. No matter who they were some of the most common words in the first few sentences are; objects, resistance, change and motion. Almost all of my recollections say something close to how physical objects resist changes in their motion. I cannot get past this type of opening statement! First off, what exactly *is* a physical object? I realize the whole standard human progression through scientific history where we all experienced and found the classical relations for everyday common objects such as a ball bearing or planet makes sense to us but this is currently just a working definition! Some always knew that we did not know whether the stuff of this place is ultimately indivisible or is infinitely divisible. What if ultimately there is no such thing as objects separate from each other in a literal way? What if *what* we really exist in is continuous? We and all objects are part of that continuity and what if what we currently perceive as objects are just the 'resolution' that we are capable of 'seeing' right now? I completely get the reasons why we have come to accept the working definition of inertia since all our classical and everyday experiences run on this and a few hundred years of hammering away in science has left us with very reproducible results. But then quantum mechanics. The very idea that objects **might** not actually exist as a 'particle type chunk of something' is right there in our face. At the moment we are still dealing with the mounting evidence for what space, time and an object is. I won't go into the quantum problems I have (me personally not the scientific community...yes I have many problems!) here but needed to throw that on the table. Resistance to change in motion ...motion? But we don't know exactly what time is or even if it actually exists independent of our memories and same thing for spatial dimensions of space itself! Motion like acceleration (which they all speak a lot about for inertia) only exists with spatial dimensions per unit time each unit of time.

Most physicists today do see that essentially everything is moving at all times in relation to something else and that finding an object which can be proven to be completely at rest to everything seems impossible.

20

So what puts or keeps all this in motion? What then really changes any of this motion? We don't know.

Energy

Energy no matter how I've tried to come at it, always ends up like other things in science … I'm using spatial dimensions, time and the concept that accelerations of *objects* allow us to '*see*' anything occur. But we don't know *for sure* what these damn things are! Another common definition is that energy is the capacity to do work … but work requires some kind of displacement against a force which typically involves an acceleration of some type (if net force is there) and a measurement of some distance. Potentials are a stored amount of energy in terms of what we know can 'come out' to be used to do work. Even in thermodynamics we were all captivated by how we could calculate the amount of energy that was 'free' for us to do work with. We categorized those pretty good and the practical applications were and are very useful. Energy in a magnetic field or electric field…or energy in a nucleus where we can 'measure' or observe the mass defect from a nuclear bomb or some such thing. But what is energy exactly? What is it not? We just don't know. Yet we have made tremendous practical progress utilizing the current working definitions/descriptions of it. We started from ideas of caloric and then managed to get to where we are today and demonstrate a lot of useful applications. I think the most common vision or thoughts about energy is like paint on a canvas or a chair in a room … most see it loosely as something that exists on or in a space. The space is commonly viewed as something that exists independent of energy….since energy is often thought of as something **in** spatial space. Alternatively, I'm sure many see or think of energy as being the space itself too. Not unlike Einstein grouped time and 3-D space into one thing called space-time. Why not space-energy or space-time-energy? The mixed message that 'loose use' of words and drawings has definitely created is how the forefront of physics sees energy/space and talks about it. Many video clips on the internet by big names in physics or movies still shows the image of the 'fabric' of space time getting bent by a clump of matter (energy) or even the fabric of space getting 'ripped'. These images and talk definitely make one think that space is some kind of existing substance or thing and then energy would be another yet. So what is the generally accepted idea for space and energy? My knowledge of physics is very thin indeed but I'm not aware of any generally taught physics even up to the 4th year university level that has any <u>substance</u> variable other than energy. Then they look at how the substance they call energy is distributed amongst the spatial dimensions which so far has been x, y and z (in textbooks...). I'm not aware of any mathematical variable for the *substance* of space itself... they use spatial variables and energy overall (time as well but the focus is spatial dimensions at the moment right here). So what is the generally accepted doctrine? Is space and energy the two substances amongst spatial dimensions?; is it just energy or something else? I myself was completely compromised all along since I did not recognize this simple choice. My mind without me even being aware always ran on a firm assumption that energy was something completely different than space (like a fabric itself of some sort). Then just recently I considered spatial dimensions to be simply an 'intrinsic' part of substance we call energy but it is information and forces that <u>*make*</u> the energy become what we CAN observe in the spatial dimensions available to it upon coupling.

Energy simply occupies the spatial dimensions it intrinsically has but both energy and its spatial dimensions become observable after coupling with I/F space. Up till thinking like this we always knew that any spatial dimensions that we observed was done through or with this physical substance (energy) thus having lead to two common perceptions;

1. Spatial dimensions independently exist like a canvas or fabric on/in which energy resides

2. Energy itself 'creates' or simply *is* these dimensions in itself.

I see neither of these. I *think* of how I/F space literally could use the substance and its available spatial dimensions as the 'tool' or the means for coupling to occur which only then leads to what we can only observe.

Some may say that is just boundary conditions. Ok... if it seems so obvious and simple...what exactly are they? I'm not suggesting my model is any better, especially since mine has plenty of simple assumptions with no justification. But we should consider what boundary conditions actually might be. Just an observation of what is in front of our face of where the energy has been allowed to sit? Why? Are we any closer to understanding why energy must do what we can all see *it* doing? (this wording is the 'current' view that most have of energy simply being all we can see and does not need I/F space coupling to yield the observable outcomes we see) And all we have to say is, 'oh...it's just the boundary conditions?...." An awful lot of careful consideration using a large amount of information must be done to come up with *appropriate* boundaries. And as you do this the energy just sits there waiting to be told what it *will* have to do or *choose* from. The substance and spatial dimensions can be viewed as possessing actual boundary conditions but having knowledge of them and what they can do and what they cannot yields the ability of I/F space coupling to make it so (observable).

Introductory post secondary physics runs on the concept of space being just x, y and z with no mention of what it might actually be. For good reason since there is no proof of anything more than these three and everything works fine when we instinctually view space like an empty room with events occurring in it. To do physics we have to have at least one substance and some spatial dimensions. How many spatial dimensions does our construct have access to? I don't know. They currently think maybe 10 plus time....but they need a substance to go with it.

What keeps energy in one manifestation versus another and what makes any of it change?

Forces

I was baffled with the whole energy and force thing. Seemingly inseparable and you can't have one without the other. But now with the idea of energy being the substance distributed amongst the available spatial dimensions of our construct, force could be viewed as the thing making it do what it is observed to do. These forces keep making the substance of our construct do what it reproducibly does and we constantly observe this. Have we ever seen force? No. Force causes an acceleration in a quantity of this construct's substance...we infer that something exists to make this occur. But we don't *see* it. Some say we can 'feel' it directly...No... you can feel the changes in the substance of this construct in its available spatial dimensions and then you assume the causation culprit is there making it happen. I'd say most are likely to be more confident in the existence of some substance in available spatial dimensions than almost anything else. Are we so sure about force? Could forces simply be the rules of the construct? Laws? Information?

Change and Distinguishability

Maybe not commonly thought of this way but I've always seen change and the ability to be distinguishable as profoundly important. And I'm not even sure why I always felt this way. But it is what it is. I do suspect that a change in the substance of this construct in its available spatial dimensions in any way shape or form is what allows us to observe or perceive time and hence change (combined with some amount of memory of it!). The ability to distinguish one arrangement from another is an absolute requirement, I would think, to acknowledge change....or to understand. Is understanding change? or Is understanding part of the change? If there was no way of distinguishing anything from another than we would have no change. Everything that we can see now in the universe appears to always be in motion relative to something and hence that in itself guarantees constant change. This is completely dependent on the forces (rules or laws) of this construct which sustains and alters what the substance is doing. If you see a chunk of iron sitting there, what is _**making**_ this clump of energy appear the way it does and stay that way for a period of time for us to see? Many simply say the laws of nature or physics but what does that mean? All the manifestations that energy has been 'forced' into provide the construct with a large number of distinguishable things or objects. Countable things. To be distinguishable is to be countable I'd imagine.

Even if space and energy were one thing intrinsically together and if all of our construct had some amount of substance (energy) to some degree everywhere we could still have objects. For a 3-D space energy construct if you had different densities of this physical stuff you could envision peaks and valleys for the densities. Even though in principle the entire thing is continuous and 'attached' you could still count the number of peaks and valleys giving the existence of things and objects. If something like this were true then we could still perceive objects (countable things) out of something continuously connected. We can see or observe objects which are more like states while what we really would like to know more about is knowledge of the *processes* between states. An understanding of why any change from one state to another took place at all. States and processes.

Information

I've definitely had my fill of searching many, many places for a scientific definition of information. There is absolutely no official and generally accepted scientific definition of information that I have been able to find. I just browsed again (to double check mainstream post secondary science) through many undergraduate physics and physical chemistry textbooks and found nothing. They all now have mentioned relativity, Quantum mechanics and some even already at least mention string theory but there is no scientific concept of information mentioned. Mathematically and computer scientifically there does exist something about information and they do draw connections to entropy for example. The biggest drive in this direction however is definitely in the areas of transmitting and storage of information that makes sense to us. Information in this sense is what we found out, discovered or even arguably created and then we try to store it and transmit it. Is that really it? Just something that can be stored and or communicated? There are growing rumblings of people talking about how fundamental information is and how it likely is something much, much more than just storing and transmitting **OUR** collections of information. What about all information beyond this? Does it exist? What is it? Some have commented on looking at whether information is just another equivalent to energy or something like that. Something that is yet another manifestation of energy. Must this all end with just energy in the available spatial dimensions it has 'access' to? Many have thought that information must have a physical representation to exist but I don't ***preferably*** see it that way. In my mind anything observable in an observable outcome space must have <u>information</u> or access to it in order to exist the way that it does.

After this brief outlined review of some of the fundamental concepts of science let's see what we now can do with this using the ideas in the existence model.

An Idea for Information/Forces?

I have never been able to comfortably sit with ideas of how a rock or animal is not as 'special' in the eternal sense as a human nor how any supernatural or metaphysical ideas are 'nonsense' with no place in science/physics. This was extremely important to me as young as 7 – 8 years old but I did not know why. Humans had supernatural and or metaphysical ideas/beliefs for thousands of years which most of the time were heavily effecting what any 'external' observer could see the outcomes of…regardless of whether they were specifically true or not. Then by the 1700's when the scientific method became more accepted and used we seen many practical and reproducible results come out of it. We also had seen the rise in how science typically keeps its distance from metaphysics. So science has become very good at working with and considering all things physically possible …which always deals with how the one physical substance (they refer to as energy) occupies the allowable spatial dimensions. The supernatural human instinct typically has dealt with something that is ***NOT*** physical or a substance specifically in our available spatial dimensions (hence supernatural). Just two weeks ago I finally seen how such a simple idea of energy being the *substance* that does what I/F space makes it do is surprisingly interesting. Not new in the minds of many physicists but nothing that warranted appearing in undergrad textbooks. I formed the ideas for a 'supernatural' I/F space over my lifetime in the form of philosophical considerations from that day as a boy I 'sensed' or just acknowledged that the truth about anything we observe and even beyond is right there but we just don't know it or can access it directly under our terms. I always said for twenty five years straight (in my head since I've thought it bad form to say it out loud) that I would know it when I seen it or found it and only then would I have a chance when my head would at last pause and rest. That has now finally been experienced. How long the rest or sense of some comfort will last? I have no idea. To allow such a track record of humans who acknowledged metaphysical ideas to go with no consideration in science is truly nonsense. No matter how much nonsense some say metaphysical thoughts are … are they not still something that is observable? Likewise for the works of science to go with no consideration if viewed as empty or soulless and cold is just as nonsensical. I realize humans are uncomfortable with this …

because that is the way they are for now. I started the philosophical section in this book with a working definition for information as I see it and I stand by it particularly because it is absolutely not a physical substance or even directly derived from it. In fact to scientifically pursue such a thing one will have to adopt this idea as a starting point. Let the majority continue to pursue what the substance can and cannot do since this is *equally* important. The scientific defining of information itself will only be achieved by those who can see it (as an idea to play with) apart from physical substance yet still as something that is very real and does interact with energy.

The idea:

Any universe has a collection of possible observable outcomes based on the fundamental substance(s) it contains and the information from information/force space that it interacts with.

In order to leave open as many considerations as possible I've worded it this way. Physics has already postulated other universes and potentially other 'types' of energy and even laws of physics but the idea sees all observable outcomes as being the interaction of I/F space with any substance/spatial dimensions (S/S) space. This then would be the sum total of all laws of physics and any other information leading to anything observable which includes thoughts and ideas. Physics to continue focusing on the physical substance is fine but the door must remain open to see if information is another player on the field of 'total' existence.

I see the concept of Substance, S and Information, I as being the two fundamental classes of what makes up everything. Each universe (let i be the number of them in total) has it's quantity of substance, S_i and information, I_i that interacts with it. Note: I specifically omitted the use of the word quantity for information because it is imperative that it is viewed as not being anything like physical substance at all and hence quantifying it in the way we do energy and all its manifestations is not *likely* (but maybe it will turn out to be quantifiable). Each universe is comprised of a number of applicable substances, s_j, where j is the number of substances. Our universe for example is currently viewed as having only one substance we call energy and thus ours has s_1 which is energy, E. The total number of observable outcomes for that universe becomes its outcome space, O_i. All universes in existence could then be summed to yield the total number of outcomes possible upon interacting with information space:

Equation 1

$$\sum_i E_i[\text{int } eract\, with]\sum_i I_i = \sum_i O_i$$

Where

Equation 2

$$s_i = \sum_j s_j$$

Our universe for example I'll call it number 1 (not for ranking purposes!) has $s_j = s_1$ since currently we explain all outcomes using just energy as our substance. Our s_1 is energy, E_1 so our equation is:

Equation 3

$$E_1[interacts\ with]\, I_1 = O_1$$

This is similar in principle with what we see in quantum mechanics where we operate on the wavefunction, ψ utilizing as always the substance of energy and spatial dimensions as required. Ψ according to one of the postulates contains all the information of the system as defined that is observable. We then get the possible observable outcomes for that system. So this idea essentially gives an **interpretation** for what we are actually already doing in quantum mechanics. What the idea also gives is the postulate of information actually being something that not only exists but is very fundamental to everything we can observe.

Use of the Model

I now can't help but use the the model to look at everything. I've chosen a few topics to try and illustrate what it can offer.

a. <u>The Quantum Model of the Atom in First Year Post Secondary</u>

I have chosen this to start with because I was utterly compelled to put my notes down and then simply talk briefly with the class this past fall. I could no longer bring myself to ***present*** essentially restatements of what is found in virtually all textbooks. Nothing with what textbooks present is 'wrong' … that is not my point. There has always been a fine line between presenting what people expect you to present and what you have come to think should be done or taught. I existed most of my life as a memorizer which consequently meant I would store exactly what was said or done with no understanding or dynamic processing of it. You could say that my processor placed a lot of this information into the 100% certain pile and thus never bothered questioning or thinking more about it! Most people I have ever met or taught are some shade of grey of this but I've always suspected that I could have been different much earlier if I could have been regularly shown the difference. This in my opinion can only be done by some who are teachers as opposed to presenters. I spent the first 8 years of my 'teaching' career knowing all along that I was a presenter of material as opposed to showing them what the learning process actually can be. So what do we present (borderline fodder for accepting as if it is truth … that is what many human processors receive it like) about the quantum model of an atom at the typical first year level?

<u>Quantum Model of Atoms</u> – some of the terms and concepts that are dropped in one chapter or even half a chapter are: electromagnetic radiation, photons, electric field, magnetic field, point of propagation, frequency, wavelength, speed of light, energy, ultraviolet catastrophe, Planck's constant, photoelectric effect, quantization, de Broglie wavelength, diffraction, interference, wave particle duality, the Bohr model, Schrodinger equation, standing waves, operators, eigenfunctions, uncertainty principle, all space, principal quantum number, angular momentum number, magnetic number, spin, spin does not exist literally but we use that word, one electron or hydrogen orbitals, electron density, collapse of the wavefunction, nodes, Pauli exclusion principle, exact solutions vs self consistent field methods, electron configuration, ….

As I was going through this standard jargon, I was literally also thinking about previous chapters in chemistry such as stoichiometry or gases where the two chapters combined could not generate that many 'sensitive to learn' words or ideas. What about physics? A whole chapter on kinematics has displacement, vectors, velocity both instant and average, Cartesian coordinate system and acceleration. How about Energy in a first year physics chapter? – work, Newtons, gravity, force, joules, acceleration, kinetic and potential energy, conservative and nonconservative forces, conservation of energy and power. And none of this is remotely as complex and least explained in beginner's words, as the half chapter of the quantum model of atoms! What am I doing? Do they (first year college/university students) have any clue what measurement for atomic structure is or was? By measurement I'm thinking of all that went into science discovering what atomic structure could be like. Has that even been clearly stated to them in a reliable and constant fashion just like order of operations is drilled into their heads since grade six? And so I stopped. Then I said, "Let's put all this completely aside for a moment. This content that the text and I have dropped on you like a ginormous boulder that even a mining excavator could not move is indisputably valuable and gained by the monumental efforts of many scientists and people. It is their work that we stand upon today but what we have come to find is that many pieces are still not fitting together as well as we would like and hence we must stay in touch with all these ideas, models, submodels, jargon that is still used but they all know it is not absolutely real or exactly true. But what at the molecular scale is essentially the only thing that all of them knew then, use today and likely will be still using for some time? We all throw tiny chunks of energy at atoms and molecules and watch to see what they do with it. Most will absorb it, 'ignore' it (a number of ways) or emit chunks of energy. Most of what we know of them is they have 'storage' for various amounts of energy that fit. Any physicist can draw two lines on the board and tell you that one is a higher energy state while the lower one is the lower energy state and the only way this thing goes from lower to higher is to absorb energy or it can go from higher to lower emitting energy. Often it is a photon that is 'caught' by the atom/molecule or is 'thrown' out by the atom/molecule. But the photon is energy... the nucleus is energy... the electrons are energy... the whole damn total of anything that physically exists is energy! So what is going on? This picture here of a photon in your book ...what is it? (They answered energy) How many dimensions spatially speaking does this picture need to 'be' the photon? (They said three) Yes, one for the direction it is moving, one for the electric field and one for the magnetic field ...what is the energy doing? (A few said oscillating) Where is it oscillating? (They were silent on this one) It is oscillating within two dimensions and propagating in one for this picture since we can only see three dimensions. So if everything is made of energy what must it be doing? (something in the three dimensions? A couple said) Has physics ever claimed there is another substance? (We don't think so) Folks, the only thing that we can do right now is measure some quantity of energy and the dimensions we think are involved. That's it. That is all we can do. So let's look again at this atom. They throw some photons of a certain energy, at an atom that they know is the amount of energy that the atom will accept. Whatever the initial shape/configuration of the 'electron density' was it is now shaped or configured differently after 'catching' the photon. That's it that we know of. If this same atom 'relaxes' it *can* (there are other relaxation routes or observable outcomes) spit out a photon of the same energy and change its electron configuration back to the way it was. The resulting observable things are what the spatial dimensions have *'allowed'* energy to use by mother nature ... or whatever/whomever controls that." Most important is for you to realize that all these are working definitions and ideas for now....We do not know any of this for absolute certain but sometimes get sloppy and almost start to believe this, that or the other thing is truth to become like a memorizing machine only..... I think I actually completely lost track of where I even was! I then looked at the class to see them and there was quite a number of them with mouths open staring at me. Then I started to panic a bit and thought, 'Oh crap, I think I went too far and irreparably confused them ... I should have stuck with the standard presentation of the chapter from the text.' Then one asked, "so they can only measure a change in energy of the atom they are 'probing'?" My answer was something like, " uhhh yes either that or something to do with where it is or how big which is tied to the other thing I mentioned which is just spatial dimensions... energy and spatial dimensions thingys". A surprising number of the class immediately said 'that makes so much more sense.' Then the largest barrage of questions and interest ensued which left me quite dazed. After class at least ten where still asking questions about stuff they've heard regarding space, time and energy etc... and 'why didn't they tell us what you did even in grade school?'. Stunning. They, just like me all those years had completely compartmentalized all this textbook stuff as if it was totally different than something as simple as how they had to take measurements in this universe! The measurement problem… but not the quantum one books talk about! In this model you can only get (not measure) an outcome….the trouble for us is interpreting it. Many at that point started to now make the connections between their measurements of energy and the ideas and models mentioned in the books. These are the ways we tried to account for what might have been happening. Can you believe it? I certainly can because that is precisely the world of misguided understanding I lived in most of my life and I do not forget what it was like…to be a memorizer with my accepted as true marching orders.

Funny isn't it… the one substance so far, energy, and what it is 'forced' to do in the dimensions allowed. But what is the force(s) or law(s) that make this so? I see it as something which is not physical, is not a substance and was here all along inches from our face … what is I/F space.

b. <u>Force and Laws of Physics</u>

I grew up thinking that force was something quite mechanical. We can see an object collide with another and become accustomed to the 'force' exerted by one on the other. Even more connected to ourselves we could not only see our hand but feel it and be part of the process when we 'will' it to exert a force on a chair for example and move it. So it is not surprising in the least that a system which has its own personal recordings of information comes to understand this concept of force they way they do. We live and breathe in a world of objects which we see as separate and we experience the concept of a 'will' to exert a force and change spatial arrangements of objects external to us. Our memory completes the process and ensures some kind of continued understanding that we have of this concept we call force. But is it real in this sense? Do we have any proof that this is correct? Not unlike the interesting collage of ideas and concepts behind the theory of the quantum atom even though all we really might know is that there are different energy states in these available spatial dimensions, so what do we know of force in a physical sense? I won't try to take us through what is said about it since there is plenty, but I will state what I think we really know. We have made our own recordings of what occurs reproducibly with the one substance we are aware of (energy) for the manifestations of it that we are aware of. That is pretty much it. We describe using *our own* information what this substance is and how much of it is involved in any observable process. So for example if a book is sitting there motionless on a table we say that we know there is a force which the table exerts matching the force of the book via gravity which is pushing down on the table. We then mention how we know the material of the table must have strong enough forces holding all of it together and not collapse under the force of the book. Gravity is a force involving mass and electromagnetic force involves something called charge. We all have heard of this along with the other two forces (strong and weak) on the current list which makes up our current recordings. Every last bit of theory surrounding force is based ultimately on substance and the spatial dimensions available, but its part is 'enforcing' the rules that substance must adhere to for the reproducibility of the observable outcomes. Problematic to that as well is that there is no mention officially as to whether these forces have anything to do with causation itself. We could argue that a mass colliding with another has 'caused' it to change its motion but we don't know what caused any initial motion or if that was indeed even the case! We do know that something is causing all we currently see to continue doing what we expect in accordance with our recordings. And yes we speak of mediators of the forces which are 'particles' in themselves (arguably) which are 'exchanged' in order to carry out the force. We have a good history as humans to prefer some kind of mechanistic interaction carrying out force for everything. It's almost instinctual. So go ahead and ask the big question … why do these force carriers do what they do? Aha! We finally end up where us humans always end up, myself included…let's see what our choices of common answers might be: god does it, mother nature does it, something does it, nothing does it, it just is what it is, the universe itself is doing it or my favorite, I don't know. I enjoy this because if I say that I'm guessing or hoping that the 'real' I/F space is responsible and I don't think that we are actually capable of gaining awareness of this not unlike how a dog cannot speak our language or understand most of what we can without even breaking a sweat; most who are listening immediately say, "but you are making up something that has no measurable proof! That is nonsense or fantasy." Then I always ask, "but do **you** know why force carrying particles are doing what they do?" They then reply, "No but at least we have evidence of these particles actually existing and appear to be involved with this force. You state that you have no proof of this actual I/F space." So now I say, "I don't have any direct proof but do you guys 'shut up and calculate'?" Now I'm seeing some fidgeting…They ask, "what do you mean?" I say, "The most 'accurate' theory you have ever used in science history (so I'm told) states that its function contains all the information of the system and often puts an asterisk on the word information indicating that it is only the observable information of the system. This does not sound quite like the collection of information that we have recorded which was historically always the case. What exactly is this information and what is this other information that is not observable?" Well now we have some silence and the fidgeting typically stops. I myself have no idea either way as to what has weight to be true and that is why I enjoy this so much since I have no personal preference inside my head. But this so called enjoyment is extremely fleeting. All I have is a terrible and insatiable desire to simply find out what the truth is. I honestly don't care about being right or wrong. Just **_try_** to get me to the truth. I absolutely agree that there has been quite the precedent over the years to give our current ideologies of force the weight is has for now. The wavefunction and Quantum 'weirdness' are really painting a picture I don't think any of us can any ignore any longer. This information and it's outcomes of what the substance must do in the dimensions available is quite compelling. It happens to be similar to what I've had in mind since I was kid and so I finally managed to envision something that is not in any way shape or form substance or physically based. Hence why it is so problematic to discover or even find. Maybe we are not supposed to be looking for it using the usual 'tricks' that I'm doubting will work any longer. We all know the saying, "God said, 'let there be light' and there was". Billions have embraced this type of answer and applied it to everything and it still is believed in to this day by many. Scientific definitions of force have only been around officially for a few hundred years (is that too much…I don't know) and they have no definite answers either. So I say, "information space holding all the information said, 'let the substance take on these properties, in these situations, in these dimensions, of this variety, …a very long and elaborate listing of information blah, blah,

blah, and humans (which is a whole other section from information space) will call it light'". And there was light.

The serious technical problem for this idea is what then is this information stuff in I/F space and how does it couple or interact to keep the rules here? I don't want to attempt to go there right now since I have no real front runner for an idea. I'll simply wait on that, but I will try to poke around in quantum weirdness to see if this idea is at least compatible for conversations' sake in a little bit. I completely agree with the scientists that until such a beast is found with enough evidence, it is not mainstream science yet. It is just an idea. But I do think that perhaps science itself should start a new category that acknowledges the actual outcomes that do occur from taking leaps of faith in theories. Yes, keep the existing journals with evidence the way they are but create a new area for publishing and accepting the potential of metaphysics. Science must contain anything that is an observable outcome.

c. <u>Spatial Energy Space Viewed as Continuous</u>

I thought about a continuous quantity of spatial dimensions and energy being intrinsically one thing. This led me to think about a number of things regarding space. Some of these ideas came up when I mentioned 'points in space'… in abstract math they don't have to account or worry about whether the infinite points on their lines or planes are real or not … but science does! So what is it? I wanted to say that each point in the space of spatial energy density must and do have a value of energy. There must be a minimum value of energy in each only because how I envision it if there cannot be a 'real' region of space without any energy nor can there be any energy without dimensions. At the same time I think about how we teach what a photon is and how the electric and magnetic fields are intrinsically bound and each is necessary to propagate it through space. Why can't spatial dimensions and energy be in the same boat in that they are inseparable? So what then is this space? Are the smallest regions separated or can it be like the mathematician's truly continuous line made up of 'points'? I pictured a contour map of land related to elevations above sea level. If your province or state had such a map with a line for every 25 feet above sea level you could then see smaller circular spots for mountain or hilltops which then truly do show countable things. But to us this landscape is a 'continuous' surface which does have countable or differentiable things just based on relative heights. Imagine even looking at an extremely flat 1 km² area but it had in the middle a 6 inch high bump ... it would be very noticeable. Do we have any concrete proof that our construct is made up of truly separate and distinct chunks of energy or matter? Clearly there are two general ideas;

1. Discontinuous Spatial Energy Space. At some point the spatial energy space itself is composed of distinct and separate pieces of itself. What separates them? What is in between them? Is there genuine absolute 'nothingness' (is this the void talk all over again?) separating them with no dimensions/substance to it at all? Is there some simple and satisfying interpretation of this that exists but we just don't understand it or know of it yet? I do not *prefer* this line of thought but I openly think of it being possible.

2. Continuous Spatial Energy Space. I prefer this line of thought for no obvious logical reason other than for the sake of finishing something written without continuously changing what I write! Maybe I see this as simpler since the whole spatial energy space is one thing all together and it can and does have distinguishability which gives us the 'illusion' of <u>*completely*</u> separate objects. If the energy density is different in one region compared to another, this <u>*does*</u> generate countable things yet is still continuous.

With that being said how does an object that we perceive move? If you seen an object moving left to right then I would say the density of the substance increases just outside it's 'current' self on the right hand side while decreasing just inside it's 'current' self on the left hand side. This change is continuously occurring. I say 'current' self simply to put in perspective that where you seen low energy density an instant ago in front of the object is now higher. If each 'point' (which likely has no meaning if it is truly continuous ...) of this spatial energy space must have a minimum value of energy or else there is no space there, then the quantity of energy is what is fluctuating like a disturbance passing but it's distribution throughout the allowable number of dimensions can also be viewed as part of the moving disturbance. So long as spatial dimensions remain populated with some minimum amount of energy then the continuity remains. I originally wrote that at least one spatial dimension would be necessary for energy to go to but I was quickly brought back to abstract mathematical thought where there is no need to worry about a physical reality and I could easily imagine in my mind all the energy existing only in one dimension or even a flat plane but I can't do that when looking out my window at a physical space. I have a hard time imagining that a truly flat xy plane of energy/substance can exist here where there literally is no z axis value (literally and actually truly zero!). This doesn't mean I'm right but I had to throw this notion on the table. I've considered it and simply prefer the 3D as the minimum (for our outcome space). Spatial dimensions higher than at least three can become active or observable by energy populating them and it is even likely to me that every dimension must have some minimum quantity of energy (again...otherwise it would not exist). This would be much like a standing baseline which would appear like a zero to those playing around with their physics measurements and such since if they can't detect or distinguish it many of them say things like, "it might as well be zero or no energy". Although this really makes me cringe since either in truth there is or is not energy there and it means something even if only in I/F space. Given the notion that I've had for some time, I definitely favor the idea that there must be energy present in a dimension or else it will not exist. Otherwise we could have the problem of dimensions popping into existence when they 'magically' get energy. For now I prefer the idea that all dimensions that can and do exist have some minimum energy in them which at such a small scale provide ample problems for finding and or proving. But as we try to approach the Planck scale experimentally speaking we should keep this in mind. It would not surprise me in the least if in fact we find that dimensions can be created by what *we* do or force the energy to do. For a quick simple example to think about ... a region of space with say 10 J of energy can undergo significant observable changes just by altering how this 10 J is distributed amongst its spatial dimensions. So for example, if the 10 J was only in three dimensions initially (again a minimum of energy in all dimensions in this model is necessary for them to be existing and can likely be there like a baseline of zero) it is a possibility we could then see 8 J 'moved' into other dimensions where we can't see how this 8 J of energy still affects the overall observable that we can see in 'our' 3 dimensions. In fact it could easily be the case that at certain scales this type of thing is not observable for us to see at all! That would be the large scales like in our everyday experience. It could be that the mm scale (or somewhere in that vicinity) and beyond there are no active dimensions altering or affecting the observables we see other than the traditional three! Hence the whole quantum world with its weirdness stuff you hear all the time about …yet no physical interpretation so just 'shut up and calculate'.

But at small enough scales these additional dimensions for energy to populate can really create some diversity. They are the factories of reality that rolls off their assembly lines adding up to the product that we and our 3 dimensions use and rely on to live the way we do.

d. Quantum Weirdness

There has been so much written and spoken of how strange quantum mechanics is. In fact in my grade 13 chemistry class when I seen the Schrodinger equation for the first time I was completely captivated. I memorized everything in school up to that point to keep all those around me happy in order to be left alone and do all the mischief I was pretty good at. For the first time in school ever I was compelled to understand how 4 numbers come out of that thing for electron configurations and to understand the 'Lego blocks' of matter. I was so heavily compromised in what I actually understood in math and science that it has taken over 20 years plus some extra help from bacteria to get to this point. It amounted to a complete overhaul of reams of stored information taken to simply be true, thus yielding a pretty crappy processor of 'new' ideas. Funny how I took what was taught in science to be true yet I thought about anything I wanted in a philosophical sense without worrying or caring if it was true or not! (And I was told multiple times by multiple people that was not science, was nonsense or metaphysical crap with some disdain) Like I thought when I was 7 or 8, the truth of anything is right here … we just don't know it. I have run as far as I can with this topic and I do not have the smoking gun evidence or theory … just a really rough idea and nothing more. This idea for a crude model though leaves no quantum weirdness at all in my mind. Truth is like magic tricks … when you don't understand how they are done or at least one of many possible understandings, you are baffled.

As biological systems that have evolved on this planet over thousands of years we have come to interpret everything around us through our senses. The number one thing we all experience and have accepted as truth is separate objects. It is so compelling and so prevalent that it is almost as if the idea of objects with separation is fundamentally hardwired into our very being. Yes there have been a few people who thought about this very carefully over the ages but that is not the norm. Continuousness versus discontinuousness ... the whole wave-particle duality. Even physics today at the most fundamental level can't help but keep talking about the new particles they have come to find along with the mass they each have. Then they speak of the non-local phenomena or the spread-out-ness of things down at the quantum level. I remember a physics professor describing to a few of us in his research lab about how force carrying particles essentially are exchanged so that they constantly 'push' the objects together or away from each other. Again instinctually particle or 'separate object based' as has always been the case for us. Just 6 weeks ago or there about I came to think of the following:

1. Spatial dimensions and time. We have always been taught through 3 spatial dimensions but if I think like a mathematician I can play around with however many is needed.
2. Physics needs at least one substance. They need some kind of physical stuff to either populate the spatial dimensions or more likely become the space. There is likely no such thing as spatial dimensions with no energy or energy without some spatial dimension. They currently have only had to use energy as their substance that I am aware of.
3. Why does that stuff do this that or the other thing? The laws of nature did it! Physics needs forces of nature to be the things that **make** the substance do what it does in the dimensions which it has available.
4. View the substance and spatial dimensions as being continuous and then view the concept of information/forces as being just one thing which does exist. This information/force is not a physical substance yet embodies all of what we think of forces doing and more.

The first three points are exactly what current physics is doing while the fourth point is the new idea for me at least. I had to realize the first three points and get them straight in my head and then play around with the fourth. After doing this all of the commonly talked about quantum weirdness just vanished. Not to say that this idea gives concrete results but it makes me see how possible interpretations can exist and that the time for shutting up, calculating and actually quietly *believing* there is no interpretation is and should be over.

The Quantum Play-Doh Model …Let's use play-doh and the various molds that kids play with at various ages as our analogy approaching in some ways how I see things. The play-doh is our substance; the molds are the forces (I/F space) that make the substance take on the shapes in the available dimensions. In the ages 3 – 6 play room at a day care the molds are simpler and easier to use so let's say we use a simple human figure mold. The kids stuff in some play-doh, close the mold and trim the excess doh off to yield the little figure. Quantum mechanically this process of stuffing the doh in would take place very quickly and the trimming of the excess would just be the extra doh being carried off while the 'molds' or information of nature make it all take place according to its rules. For the quantum play doh we could then even model photons interacting with atoms. Any change in the shape of the little human figure has to have the exact amount of play-doh thrown at it, the speed of which is always the same …the mould itself is quite malleable in this quantum play doh world and if you throw the exact amount of doh that matches the amount needed to get to the next shape of the human figure then it can happen, otherwise the doh you threw at it simply just keeps moving. If there is a match its absorption occurs instantly without any slowing, almost as if both the human figure and the extra piece of doh thrown 'knew' there was a match. Let's say in our case the figure simply gets slightly wider or taller, but there is a noticeable difference. So now let's say that the kids threw every possible amount of doh at their human figure and mapped out the only absorptions and hence resulting shapes and sizes it can take so they feel pretty confident they have it all figured out ….so they then wander over into the ages 7 – 9 room and repeat the 'mapping' exercise but notice some new absorptions. The human figure is essentially the same but it now can have a play-doh hat on it. This one figure can only be with or without a little tiny hat, while others can have two hats, 7 hats or some can have none. The absorption of quantum doh is an exact amount to yield each of the hats in each case. Then they realize that changing the local environment changes the 'molds' of the system in their midst and they have to remap out all the possibilities and record them so we know. But information space already knows of all the molds …it *is* all the molds for that substance! Let's just say that this ages 7 – 9 room was close to a moderate magnetic field. Oh!!!! When you put a play doh figure into a magnetic field some get new absorptions of little bits of doh while others do not. The molds can change. The forces/information of this nature make it so. The play-doh is very dynamic and malleable and it simply is forced to become whatever the person molds it to be in the available dimensions they can access with the molds they can use. (It could be that forces are not even necessary…only information and outcomes) The doh spreads itself out in the mold the way it is forced to. In this quantum play-doh model though each child cannot simply mold the doh into whatever they want … let's say they have to use available molds which the day care has put into use at that time and possibly 'forever'. Thus the objects you will ever see there are one of many off the list of what can be. These kids can think and dream of all sorts of fantastic shapes, extra dimensions and such but many of this cannot be. It is forbidden… it cannot exist here at all. But there are other Day Care facilities that have their own rules and molds … and then there is the one place with no doh at all but it contains an infinite array of molds! This is where all molds come from and Day Care facilities all must 'rent' their molds from this place. In fact this is where all molds come from and where anything a child can even think of already 'exists' *informationally* there! If the kids play their cards right they might be able to discover what other molds they can get and use.

So now back to our quantum world …

Quantum mechanical spin is no longer strange at all. The only thing strange is the road we took to learn about it being there! Physics today has already essentially completely agreed that there is no literal thing such as an electron or nucleus spinning in a classical sense. The only thing they ever noticed was that when you put some things in a magnetic field you get more or different absorptions of energy… that's essentially it. Changes in distinguishability. They noticed something different in one or all of these:

- the amount of energy in the system

-   the distribution of energy in the system (which is often a change in the shape of the system or even something in it)
-   the behavior of the energy we threw at it such as reflecting or scattering etc…

So when the local environment of our atoms changes like putting it in a magnetic field the laws of nature put a new mold in place to make that system be what it is or could be, since some random choice might be very real. Physicists would say something like, "the forces which apply at that level and in that situation make it so". Either way the substance which we think of as energy must do one of the three things just outlined and even do it in the dimensions of the 'new' mold. There is no spin. All we have witnessed is a distinguishable thing the substance can do in our universe according to I/F space. In a magnetic field what used to be degenerate energy 'levels' are now distinguishable. Why? Because the laws of physics made it so and we do not know the answer yet. It is all about the rules or laws which is this information.

The two slit experiment? There should not be an interpretation problem. Whether the one I'm thinking of is actually correct is not the point to me … it is the fact that the only thing I have faith in is that the truth of what it is, is what it actually is!…And we will find it, try to find it or simply keep making our recordings for our records get better at approaching it. First off in this model there is no separate object or particle heading towards the two slits. If you are someone who has read this far and are not familiar with it (classic two slit experiment) please take a quick look at numerous videos online which are quite good. One in particular that is often used today is 'Dr. Quantum' animated videos which are great. Have a look then proceed. If we shoot an electron at the slits all we have is a region of space with an energy density (in however many dimensions it actually is allowed to use) that is different than its local surroundings making it distinguishable to us … it's object-ness comes from this distinguishability. This then interacts with the slits which is again a region with its own energy density and attributes. The laws or molds of nature then make our 'electron' change its energy density into whatever it exactly is just after the slits yet not at the screen. We call this a state of superposition which has no agreed upon interpretation. In this model the energy of that electron due to the particular interaction it had at the slits makes its energy spread out the way it does in however many dimensions it can or does because that is exactly what it is allowed to do. Energy at this scale is very dynamic and malleable and this electron does not break up like a rock into multiple 'pieces' which then interfere with itself! The model sees this as a decoupling of information and substance which is essentially the absence of an actual observable outcome for the time of the decoupling. But yes the energy density (substance) of that electron is simply spreading out the way it can over the space it can but the most interesting part is that the 'information' of this electron is not lost either since it is 'from' I/F space. The 'rules' or characteristics of this coupling/decoupling process are what they are and we are learning more about them all the time. Certain things having coupling take place to then 'bring' back the electron for us at the screen as an observable outcome. The only difference we know of is that its location must be in one of the regions it is allowed to be in and many are not the classical trajectory. The energy didn't 'disappear', the information describing it didn't go anywhere but the distribution of it in the mold that nature puts it into after such a two slit interaction is simply what it is. You never had a separate isolated particle to start with! Many call this return to being an electron at the screen as the 'collapse' of the wavefunction but this model simply says decoupled.

Quantum tunneling? Sounds mysterious but it is not and the math essentially already has an interpretation for it but we don't…because the object does not exist the way we always thought in the first place. Just like the classic aluminum oxide layer that should not let electrons flow 'through' it. The classic physics model(s) say that can't be done period but we observe it occur all the time. The quantum wavefunction for this electron has plenty of values on the other side of the oxide layer and we call this tunneling. But if the electron never was a 'separate' distinct item but was simply a region of space with a particular energy density which spreads out according to whatever the mold of nature is under those circumstances then the 'barrier' or wall isn't there either like we always thought. It does not 'tunnel', it simply gets redistributed according to the probabilities that go with it or makes the density shift to where it can and does. The distribution of energy in every single 'point' in the entire universe is what it is at any point in time and the overall mold is instantaneously what it is according to I/F space. Whether we trap a small quantity of energy in a theoretical 1-D box and show all this wonderful quantum stuff or we have a large amount of energy in the universe's box it's the same concept but just more complex. Everything we observe in it is just what the energy does spot by spot according to the information of nature. Say that it does what the forces of nature make it do if it feels better...but physics is looking and many believe there must be a unified theory of everything which they often say, "there is likely a unified single force which includes all the others we use right now". Quantum already is giving us the very first tools to see or infer that information is something that is real and is the fundamental 'force space' providing the descriptions of how energy must behave under the local conditions we find it in at the small scale and it does 'add up' or culminate in the large scale structure too.

Stationary state. So far I've only talked about this in a semi-stationary way since I have not actually talked about time dependence too much. Inertia is a huge item and even in this small scale world there must also be constant motion or movement of something otherwise up at the larger scale why the whole inertia thing? Energy or energy densities of all the regions of spatial energy space are constantly redistributing (decoupling/coupling) all the time. Even in an atom such as hydrogen's ground state with its 1s electron or the contents of its nucleus like the proton or the quarks or whatever else is down there is all in motion of some kind. Typically we would likely favor some kind of fluctuation in the value of the energy density within the small region of the electron's orbital space for example. The energy density is always fluctuating or redistributing in time and if we model a situation where we 'leave' the hydrogen come to an equilibrium we get a 'standing' wave or should say a repeating energy density fluctuation in time which if plotted or had a picture of would look like a wavy thing. That's why the ring of iron atoms will have 'standing' energy density waves in its picture. So after all the years of how we've been taught to see this construct we have immense trouble clearing out thousands of years of seeing everything as distinct separate objects or 'particles' in this case and it *likely* could be wrong. Yes I agree with some physicists that have said, "Mike we know they can't be 'real actual particles' down there since the evidence is quite obvious that they are not, but we have to refer to specific regions of energy as something since we can see this is an electron or this is a proton. We know they are not specific distinct objects!" The number of scientists that clearly see this and actually think of them this way is exceedingly small. In fact because nature does indeed localize an electron for instance when we take a measurement to something that really seems to be a definite particle many are still very skeptical as they should be. All I see though is this collapse in concert with I/F space is likely to be instantaneous and the immediate return to spreading the energy density back out to what it can follows. From our side of the fence time is very real while on the other side in I/F or S/S space there is no time like we have it here. Whatever the list of interactions are that cause the collapse to a 'point' or very tight spot of energy density, is what it is and we are finding out more and more about this list. Physicists are getting better at keeping the superposition up and running in the time we hold it and the size. They are also getting better at finding out what causes it to collapse and what does not. They are learning more about the 'molding' of energy density that information (or the forces of nature) already knows and uses.

Quantum Teleportation. I know very little about this so I'll just move ahead with the little I have read. My understanding is that they have found a way to send a large percent of a message from one location (A) to another (B) 'instantly' or at least they all agree that it was clearly faster than the speed of light but they cannot 'read' or comprehend this message unless they send 'the key' or the remaining percent of the message via the standard speed of light method. Sure it sounds strange and all that but not anymore to me. The two slit experiment makes me think that the spreading out of energy density and the fact that the sheer size of this spread can in fact be over the entire universe but reined in by the amount (or probability values) in these remote places. Interesting though, how in a more contained superposition space like between the slits and the screen the 'electron' has spread out primarily in it but a particular interaction brings it all (including its information that makes sense to us) back to a much more localized point or region again ... so why does the teleportation thing seem so strange? I see elements of the same thing operating there. Are the remaining 'pieces' of information about the energy density we were dealing with somehow tied up at location A in what we can only send via light methods to the other location B where the rest of the 'collapse' information in that energy is 'waiting'? Some pieces of information can cover the distance instantly while others cannot ... for now with what we know of. Are we not just learning something else about what this substance in the dimensions it operates in can do? Add it to the growing list of what we learn about it … this list will just get closer and closer to what I/F space already knows exactly…and that is just the truth for which we all are in search of. I need to restate what I think I've mentioned in this book somewhere but I can't even remember to go to a lecture sometimes so I'm sorry about this. Information is a funny thing since it is not a substance … it does not behave like energy in spatial dimensions or time in the same way at all. Information and 'the key' like in the teleportation experiment … some things have a 'sense' about them like they have already happened but not at the same time! We call this weird but I no longer see it that way at all. I/F space does not have time or spatial dimensions nor its own consciousness … but it is responsible for 'molding' all we observe with an element of randomness/gaps in info. This book for example … I don't think it or feel it but there is a strange 'sense' about it (whatever that means!...but I'm stuck with human words here) that the someone or some small group will be *or* is already 'found' by it! From our perspective we have to say the book will find them because we do experience time … but it is already 'done' but not done at the same time in I/F space! Think about the interference pattern behind the slits where the electron might think it could decide to be found directly on the screen where the actual 'node' is but in fact it never will (provided the environment stays that same way…). Feelings, hopes, dreams or ideas of what could be are not the same as what can actually be. In this construct that electron 'might be able' to choose without knowing it from a constrained list of outcomes whilst 'dream' of many others it can see or hope for, but that is not possible for it. Its list of most probable outcomes is what it is but it just does not know of it. As a human you might feel so afraid of getting cancer before 50 like many others in my family and that might very well be true but it also could be true that it is not on your actual outcome list…what are you 'superimposing' for? … but sometimes people have a strange sense of what impending fate exists for them… this of course drags us into metaphysics and mystical supernatural stuff as always and has always been part of human history. I don't subscribe to this type of stuff but I do subscribe to the slightest detail in the lab and refuse to neglect any idea that could be tied to something true… and metaphysical ideas do occur. Funny how I am terrible with details but yet certain ones 'find' me. Everything that is observable has a meaning, even if 99.999% of it, especially after receiving it through human conversation, is science fiction fantasy. Animals have fantastic detectors. Even the deep sea animals that have a detector for the faint electric fields of other animals nearby! Even the idea that our eyes have been rumored to be able to detect single photons is interesting in itself. I don't ascribe supernatural stuff to this at all but I am intrigued about the evolution of this universe from hot quark soup, to forming atoms to forming single cell machines etc….I/F space, this construct and the sequence of outcomes that have occurred and are occurring right now is science. The model you use or run with can 'choose' to exclude much, but that is typically an effort to simplify what we humans refer to as 'too complex for us to handle' with any comfort …a simple model is 'good enough'. 'Good enough' might be good enough for humans but I/F space does not use a model to make its existence bearable.

e. <u>Entropy</u>

I think it is very important to go over the mathematical concept of entropy first before talking about why I think it is so important. In high school we are taught to memorize a definition such as: 'entropy is disorder or randomness' and 'the universe is constantly heading to a state of disorder'. These types of words accompanying entropy caused a lot of trouble for me since I had a very different fundamental definition of disorder in my head at a young age but we'll get to that. Pretend this is the first time you have ever learned of entropy and lets approach it by a stats class path.

Let's say we have 4 baskets which for starters can only hold one item at any point in time. 2 have 1 apple and the other 2 have 1 pear in them. apple = a and pear = p. I'll put in my hand written notes since I don't have the patience to type all this crap in using equation editor or whatever it is called.

4 baskets holding
2 apples and 2 pears

This is the state of our 4 basket system so if we were blindfolded and these were mixed up and we reached into 1 basket what are the chances we would grab an apple? Well we know this set of 4 baskets had 2 apples and 2 pears so... the chance or probability of getting an apple (p_a) is $\frac{2}{4}$ or 0.5 and probability of getting a pear (p_p) is also $\frac{2}{4}$ or 0.5.

Does "order" mean anything?

$\Downarrow$

i.e.
i) a a p p
ii) a p a p
iii) a p p a
iv) p a p a

$\vdots$

$\downarrow$

Ordering this differently does not give a different S!

To be clear ... look at the state of your system, then imagine randomly grabbing one of the baskets asking yourself, "Knowing the state was 2 apples & 2 pears what is the probability of grabbing an apple or a pear?". You need the probabilities of grabbing the things in your system!

Entropy, $S = -\sum_i p_i \ln p_i$ $\Rightarrow$ this is the stats class definition for entropy! p_i is the probability of "finding" the ith thing in your system you know

So for our a a p p system.

$S = -\left[p_a \ln p_a + p_p \ln p_p \right]$

$= -\left[.5 \ln .5 + .5 \ln .5 \right]$

$= -\left[-0.3466 \quad (-0.3466) \right]$

$= 0.6931$

this entropy is the same for a a p p, a p a p, a p p a, p a p a . etc because the probabilities stay same!

So what if we had .....

a a a a   or   a p p p

↓

blindfold yourself and
reach in ... what are
the chances of grabbing
an apple? ⟹ 100% or $p_a = 1$

$S = - \sum_i p_i \ln p_i$

$= - p_a \ln p_a$ (there is no
other item in
these baskets)

$= - 1 \ln 1$

$S = 0$ (pppp would also
have S=0)

— blindfold
— reach in and ask, "what is $p_{apple} = ?$ + $p_{pear} = ?$

$p_a = \dfrac{1}{4}$; $p_p = \dfrac{3}{4} = 0.75$
$\quad = 0.25$

$S = - \sum_i p_i \ln p_i$

$= - \left[p_a \ln p_a + p_p \ln p_p \right]$

$= - \left[0.25 \ln 0.25 + 0.75 \ln 0.75 \right]$

$= - \left[-0.347 + (-0.216) \right]$

$= 0.563$ (paaa would also have
$S = 0.563$)

Summary:

Situation of two types of fruit (apples + pears) in **4** baskets	Entropy, S
a a a a	0 } lowest
p p p p	0
a a p p	0.6931 (highest)
a p p p	0.563
p a a a	0.563

How high can we make entropy go with 4 baskets and 1 item per basket? One different fruit in each basket. So lets add an orange, o and a bannana, b. $\boxed{a}$ $\boxed{p}$ $\boxed{o}$ $\boxed{b}$ $\equiv$ a p o b   — blindfold
— randomly reach in
— prob for an apple, a pear, an orange or banana?

$$S = -\sum_i \rho_i \ln \rho_i$$

$$= -\left[\rho_a \ln \rho_a + \rho_p \ln \rho_p + \rho_o \ln \rho_o + \rho_b \ln \rho_b \right]$$

$$\boxed{\rho_a = \tfrac{1}{4} \; ; \rho_p = \tfrac{1}{4} \; ; \rho_o = \tfrac{1}{4} \; ; \rho_b = \tfrac{1}{4}}$$

$$= -\left[0.25 \ln 0.25 + 0.25 \ln 0.25 + 0.25 \ln 0.25 + 0.25 \ln 0.25 \right]$$

$$= -\left[0.25 \ln (0.25)^4 \right] \quad \text{or} \quad -\left[\tfrac{1}{4} \ln (\tfrac{1}{4})^4 \right] \text{ same}$$

$$S = 1.386$$

⋔ NOTE how all "outcomes" of you reaching in to <u>that</u> situation of baskets are equal probabilities!

ie $\rho_a = \rho_p = \rho_o = \rho_b = \tfrac{1}{4}$

<u>This</u> gives largest value for entropy with a simpler formula.

how about...

b b o p? $\rho_b = \tfrac{2}{4}, \rho_o = \tfrac{1}{4}, \rho_p = \tfrac{1}{4}$

$$S = -\left[.5 \ln .5 + .25 \ln .25 + .25 \ln .25 \right]$$

$$= 1.04$$

So note again the equal probabilities thing.... (we'll have kiwi too! = k)

2 baskets ⇒ ob $S = -\left[\tfrac{1}{2} \ln \tfrac{1}{2} + \tfrac{1}{2} \ln \tfrac{1}{2} \right]$

orange, banana

$\rho_o = \rho_b = \tfrac{1}{2}$

$$= -\left[\tfrac{1}{2} \ln (\tfrac{1}{2})^2 \right] \quad \cdots$$

recall algebra:

$\ln A + \ln B = \ln AB$

$$= -\left[2(\tfrac{1}{2}) \ln (\tfrac{1}{2}) \right]$$

and $\ln A^x = x \ln a$

$$S = \ln 2$$

$$S = 0.693$$

and $-\ln A = \ln \tfrac{1}{A}$

3 baskets with equal probability outcomes . . . (orange, banana, apple) $P_o = P_b = P_a = \frac{1}{3}$

o b a $S = -[P_o \ln P_o + P_b \ln P_b + P_a \ln P_a]$

$$= - \left[\frac{1}{3} \ln \frac{1}{3} + \frac{1}{3} \ln \frac{1}{3} + \frac{1}{3} \ln \frac{1}{3} \right]$$

$$= - \left[\frac{1}{3} \ln \left(\frac{1}{3}\right)^3 \right]$$

$$= - \left[3 \left(\frac{1}{3}\right) \ln \frac{1}{3} \right]$$

$$= - \ln \frac{1}{3}$$

$$S = \ln 3$$

Guess what happens for 4 baskets (o, b, a, p) and then 5 baskets (o, b, a, p, k)

$$\left(P_o = P_b = P_a = P_p = \frac{1}{4} \right) \qquad \left(P_o = P_b = P_a = P_p = P_k = \frac{1}{5} \right) \text{kiwi}$$

$$S = \ln 4 \qquad\qquad S = \ln 5$$

So if each probability is equal we get the largest entropy! and the formula goes from $S = \sum_i P_i \ln P_i$ to $S = \ln \Omega$ where Ω is # of equal probabilities or

some say $S = \ln N$ where N is the # of different items of equal prob.

So . . . if you had 4.5 million baskets each with a different item

it's entropy = $\ln (4.5 \text{ million})$

if you had 4.34×10^{21} Hydrogen atoms where each was in a different energy state . . . .

$$S = \ln \left(4.34 \times 10^{21} \right)$$

and so Boltzmann put in his constant k_b

$$S = k_b \ln \left(4.34 \times 10^{21} \right)$$

$$= 1.381 \times 10^{-23} \frac{J}{K} \left(\ln 4.34 \times 10^{21} \right) = 6.88 \times 10^{-22} \frac{J}{K}$$

So what if $\frac{1}{2}$ of 4.34×10^{21} Hydrogen atoms were in $\underline{1}$ energy state(A) while the other half was in another energy (B) state? This would be identical to having just $\underline{2}$ hydrogen atom where 1 was in the one energy state and the other was in the other energy state!

# in state A	# in state B	Entropy
1	1	$S = k_b \ln 2 = -k_b \left[P_A \ln P_B + P_B \ln P_B \right]$
2.17×10^{21}	2.17×10^{21}	$S = k_b \ln 2 = -k_b \left[P_A \ln P_A + P_B \ln P_B \right]$ $= 9.57 \times 10^{-24} \frac{J}{k}$ Entropy is calculated by the probabilities of when you randomly reach in to the system to get something. Here it's 50:50!

Now • we're getting somewhere.... ▪ psychological "studies" or pictures they show then ask things like, "What do you see? ... what do you think of?". Or "I'll hold up pictures and I want you to tell me if you think of order or disorder"....

1st picture: | 1 2 3 4 5 6 7 8 9 10 | ⇒ Everyone says order

It's the same entropy!

2nd picture: | 4 10 5 7 2 9 6 3 8 1 | ⇒ Everyone says disorder

But any group of ten different things has same entropy because we have 10 equal probabilities! Yes we have the greatest variety of things possible but to say disorder and random mess is subject to a lot of interpretations.

So where did the word "randomness" or disorder come from for the topic of entropy?

One way to look at this is a box with $\underline{4}$ fruit...

4 apples   vs.   4 different fruit

$S = 0$ $S = 1.386$

all same is more "uniform" or "ordered" or "clean"

all different is "less uniform" or "more disordered" or "messy"

→ note how the use of words here is not talking so much about sequencing "order" but actually associating "variety" with disorder.

→ this problem with meanings/understandings of words can be very problematic especially for memorizing style processors, but calculating the actual number for entropy of a system only requires you to $\underline{assign}$ probabilities!

another note: How and where do we draw distinguishability "lines" for a system?

ex.

(A)

1 box, 4 fruits only "one" location (just in the box)

(B)   .   (C)

1 box $\underline{with}$ 16 holders.... 4 fruits. If when you reach in you only check $\frac{1}{16}$ holders the P_i change vs situation (A)

I'll stop travelling further into the entropy "hole". I am no expert so far this simply provides some good basic ideas for the entropy concept.

So statistically:

$$S = -\sum_i p_i \ln p_i \quad \text{this is the foundation.}$$

Gibbs/Boltzmann

$$S = \left(-k_b \sum_i p_i \ln p_i\right) \qquad S = k_b \ln \Omega \quad \text{or} \quad k \ln N$$

if system
as all "N" different items with equal probabilities

Traditional Entropy in High School (thermodynamics—macro)

$$S = \frac{q}{T} = \frac{heat}{temperature}$$

→ turns out that at molecular level energy gets distributed amongst all the molecules. This is like distributing things amongst baskets and all systems end up maximizing entropy at equilibrium where probabilities are equal.

— This redistributing or changes from state to state is done by random collisions like diffusion <u>not</u> entropy itself! Entropy is only a value of the <u>state</u> not the transition driver. That is the "forces" or laws or rules of the physical construct.

Let's consider 4 boxes; 4 marbles which collide in pairs. The marbles are always in boxes and 1 marble must be exchanged per collision. All "decisions" are made randomly. Start with a 1,1,1,1 state.

So boxes a b c d each have.

1 marble ⟶ 1 1 1 1

ex collision between box a and b ⇒ ab

outcomes
i) a gives b 1 marble
or
ii) b gives a 1 marble.

so...1111 $\overset{ab}{\diagup\diagdown}$ 3011
  0211

Collision Scheme ①

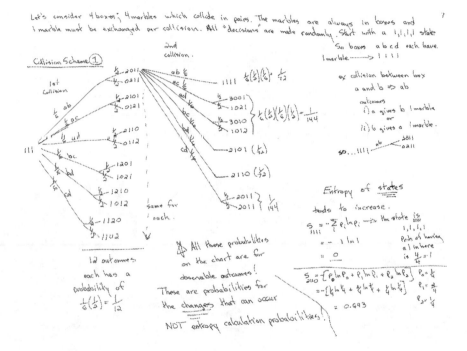

$\frac{1}{2}$ 2011
$\frac{1}{2}$ 0211

$\frac{1}{2}$ 2101
$\frac{1}{2}$ 0121

$\frac{1}{2}$ 2110
$\frac{1}{2}$ 0112

$\frac{1}{2}$ 1201
$\frac{1}{2}$ 1021

$\frac{1}{2}$ 1210
$\frac{1}{2}$ 1012

$\frac{1}{2}$ 1120
$\frac{1}{2}$ 1102

1st collision

$\frac{1}{6}$ ab
$\frac{1}{6}$ ac
$\frac{1}{6}$ ad
$\frac{1}{6}$ bc
$\frac{1}{6}$ bd
$\frac{1}{6}$ cd

ab $\frac{1}{2}$
ac $\frac{1}{2}$
ad $\frac{1}{2}$
bc $\frac{1}{2}$
cd $\frac{1}{2}$

1111 $\frac{1}{6}\left(\frac{1}{2}\right)\left(\frac{1}{6}\right)^2 = \frac{1}{72}$

$\frac{1}{2}$ 3001
$\frac{1}{2}$ 1021
$\frac{1}{2}$ 3010
$\frac{1}{2}$ 1012   } $\frac{1}{6}\left(\frac{1}{6}\right)\left(\frac{1}{6}\right)\left(\frac{1}{2}\right) = \frac{1}{144}$

2101 $\left(\frac{1}{12}\right)$

2110 $\left(\frac{1}{12}\right)$

$\frac{1}{2}$ 2011
$\frac{1}{2}$ 2011 } $\frac{1}{144}$

same for each.

12 outcomes each has a probability of $\frac{1}{6}\left(\frac{1}{2}\right) = \frac{1}{12}$

∮ All these probabilities on the chart are for observable outcomes! These are probabilities for the changes that can occur NOT entropy calculation probabilities!

Entropy of states

tends to increase.

$S = -\sum_i \rho_i \ln \rho_i$ ⟶ the state is $\frac{1,1,1,1}{1111}$

$= -1 \ln 1$ Prob of having a 1 in here is $\frac{4}{4} = 1$

$= 0$

$S_{2110} = -\left\{ \rho_0 \ln \rho_0 + \rho_1 \ln \rho_1 + \rho_2 \ln \rho_2 \right\}$ $\rho_0 = \frac{1}{4}$

$= -\left\{ \frac{1}{4} \ln \frac{1}{4} + \frac{1}{4} \ln \frac{1}{4} + \frac{1}{4} \ln \frac{1}{4} \right\}$ $\rho_1 = \frac{3}{4}$

$= 0.693$ $\rho_2 = \frac{1}{4}$

Now let's look at 10 atoms and 20J of energy. Each atom has 10 different energy levels 1J, 2J, 3J - - -> 10J. It's spread out this way to keep it simple for now.

The lowest entropy STATE for our 10 atom system to distribute 20J of energy is: 2,2,2,2,2,2,2,2,2,2 (all 10 atoms in their 2nd energy state of 2J

I'll use check marks to show each atom's energy state. (~✓ ✓ ✓ ● ● ● ● ✓ which check is proper? left or right? (sinister) (dextr.

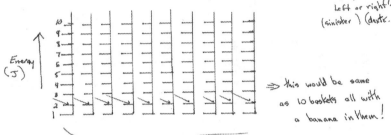

Energy (J)

⇒ this would be same as 10 baskets all with a banana in them!

10 atoms

- blinfold yourself
- grab at random 1 of these atoms
- what are the chances of 1 of them being in it's 3rd energy state? ⇒ none cause we know none are in that state for our overall system's state. $P_3 = \frac{1}{10}$ finding 1 in 2nd energy state

$$S = -\sum_i P_i \ln P_i$$

$$= -\left[P_2 \ln P_2 \right]$$

$$= -\left[1 \ln\left(\frac{10}{10}\right) \right]$$

$$= 0$$

even if we had ● 2.30×10^{15} of them S=0 if they were all in same state...whether all we in 1st, 2nd, 3rd ... BUT how much energy the system has determines what configuration it can exist in.

So can this system exist where all are in 7th energy state? ⇒ No. cause that would require the overall system to have 70J of energy!

for 10 baskets to have maximum entropy all we needed was 10 baskets each with a different fruit in it. So for our atoms we would need...

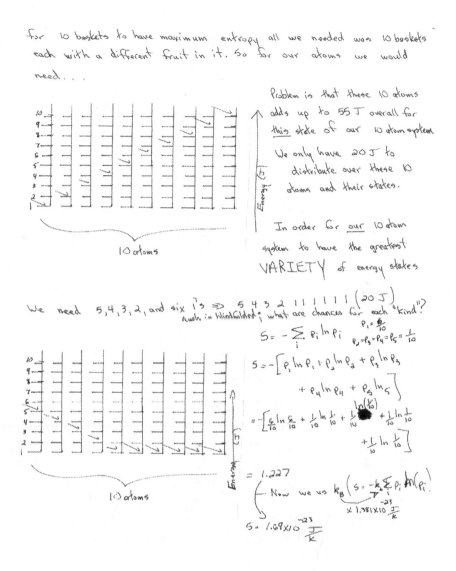

Problem is that these 10 atoms adds up to 55 J overall for this state of our 10 atom system

We only have 20 J to distribute over these 10 atoms and their states.

In order for our 10 atom system to have the greatest VARIETY of energy states

We need 5,4,3,2, and six 1's ⟹ 5 4 3 2 1 1 1 1 1 1 (20 J)
(walk in blindfolded) what are chances for each "kind"?

$$S = -\sum_i p_i \ln p_i$$

$p_1 = \frac{6}{10}$

$p_2 = p_3 = p_4 = p_5 = \frac{1}{10}$

$$S = -\left[p_1 \ln p_1 + p_2 \ln p_2 + p_3 \ln p_3 + p_4 \ln p_4 + p_5 \ln p_5 \right]$$

$$= -\left[\frac{6}{10}\ln\frac{6}{10} + \frac{1}{10}\ln\frac{1}{10} + \frac{1}{10}\ln(\frac{1}{10}) + \frac{1}{10}\ln\frac{1}{10} + \frac{1}{10}\ln\frac{1}{10} \right]$$

$= 1.227$

Now we us k_B $\left(S = -k \sum_i p_i \ln(p_i) \right)$

$\times 1.381 \times 10^{-23} \frac{J}{k}$

$S = 1.69 \times 10^{-23} \frac{J}{k}$

These previous 9 pages of my handwritten notes strays from what I really wanted to say about entropy but there is no way I could pull this off without making sure we all are on the same page with the concept of entropy. It is not all about disorder when the majority of human understandings of the word does not match properly. Consider the following picture of picnic baskets scenarios:

Michael Siwek

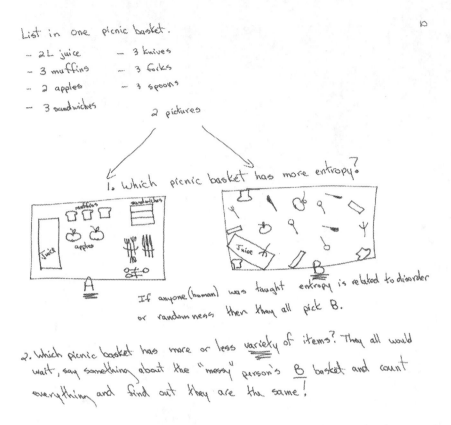

List in one picnic basket.

- 2L juice
- 3 muffins
- 2 apples
- 3 sandwiches
- 3 knives
- 3 forks
- 3 spoons

2 pictures

1. Which picnic basket has more entropy?

If anyone (human) was taught entropy is related to disorder or randomness then they all pick B.

2. Which picnic basket has more or less variety of items? They all would wait, say something about the "messy" person's B basket and count everything and find out they are the same!

The entropy of both baskets is the same! Stop relating the _commonly_ used word disorder for entropy when teaching it. Consider what I was teaching for years because I _believed_ it was correct … the figure of two rooms across a hall where as soon as one person gets a ball they throw it into the other room was a popular thing I _presented_ to the class. At equilibrium the rate of 1 ball going one way equals the rate of a ball going the other way.

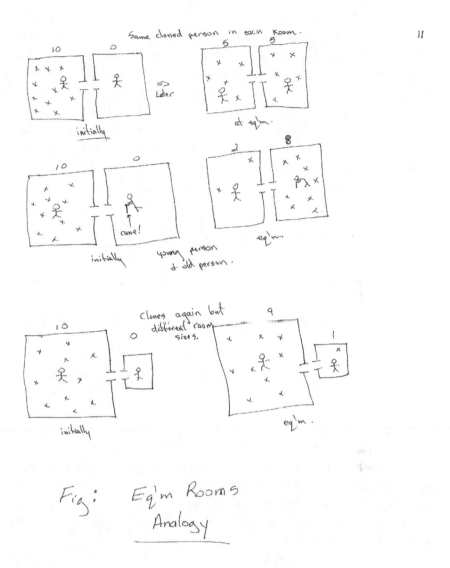

Same cloned person in each Room.

II

initially ⇒ Later at eq'm.

initially came! young person + old person. eq'm.

Clones again but different room size 9. initially eq'm.

Fig: Eq'm Rooms Analogy

With two identical people (clones) and room sizes, the ratio at equilibrium is easy to see should be 5:5. Make one old and the other young in the same size rooms and the ratio shifts as shown. Greatly alter the size of one of the rooms compared to the other even if both are clones you get a shifted ratio again. Why? Entropy is what I thought was at the root of this! For the bigger room this is less organized or messier/ disordered because the balls are spread out in such a big space while in the very small room they are kept organized or ordered and hence less spread out. For the older person who moves so slowly this might as well be the same as making the room huge with a mess! Entropy is a measure of a state not the actual process in time! This is just like I mentioned with the collision chart above. Altering the size of room or abilities of the persons is just like altering the rules of laws of collisions in this chart. This is not entropy! But which room scenario has the highest entropy and which the lowest? Takes some time to think about it if disorder is in your mind by our common understanding of that word doesn't it. Calculate it! (stats entropy with no units … no k_B constant)

12

$$S_{5:5} = -\left[P_5 \ln P_5 \right] = -\left[\frac{10}{10} \ln \frac{10}{10} \right] = 0$$

$$S_{2:8} = -\left[P_2 \ln P_2 + P_8 \ln P_8 \right] = -\left[\frac{2}{10} \ln \frac{2}{10} + \frac{8}{10} \ln \frac{8}{10} \right] = 0.501$$

$$S_{9:1} = -\left[P_9 \ln P_9 + P_1 \ln P_1 \right] = -\left[\frac{9}{10} \ln \frac{9}{10} + \frac{1}{10} \ln \frac{1}{10} \right] = 0.325$$

what about

$$S_{6:4} = ? \Rightarrow 1.222$$ we must approach equal probabilities of different things.

Consider a class of 23 students for a test out of 37 marks. If they all got 30/37 then the entropy for that state of marks is 0. If none of the 23 student's marks out of 37 were the same then the entropy for that state of marks is at a maximum of 3.135. Teachers or people using statistics don't normally ask, "what was the entropy of your grades?" … what they do ask is something like, "what variety or spread of marks did they get?". If you say your grades had order to them that leads many to think of the sequence of the marks… careful… if you had 100 students and each populated 0% up to 100% right on the spot then that is a perfect singularly occupied sequence in order! But that is higher entropy than having the 100 marks spread between the sequence of 20 % to 80% since many of these marks are duplicated despite the ordering of all grades between 20 and 80%.

Disorder. This has the word order in it and most of us automatically think of sequences or the ordering of something. Entropy does not care about sequences or the order of the things ... it cares only about the largest different number of things in a collection. The greatest distinguishability. Perhaps a new definition when using just words could be something like:

Entropy: is a measure of the variety of different things in a collection of countable things by their fraction of occurring in that state.

I debated whether or not to include the last 14 pages or so since it has nothing directly to do with what I was most interested in entropy for which is information/forces of nature which make energy do what we see it do! Trouble was that I *believed* entropy was responsible for altering the probabilities in an outcome chart as we move from one state to another! At times I actually was convinced that entropy and information should be either the same thing or directly related. Any of you that fully get entropy already should be surprised why I 'wasted' so many pages to talk about something most SHOULD already know. But my own quick survey so to speak showed me that overwhelmingly most people from high schools, colleges and universities do not really get what was covered here so far for entropy! **I didn't** and I was supposed to teach the stuff! One of the classic misuses (in my opinion) of entropy was in thermodynamics where people say that energy is unavailable because it gets, 'tied up in entropy' of the system. I completely agreed because I thought entropy was responsible for energy *getting scattered* in the various _processes_ just like the reasons behind why the two rooms/balls/hallway scenario played out. But it is not. Only processes and the rules/forces/information that guide them can get a system from one state of entropy to another state of entropy. Not entropy itself. Entropy by itself does not force a system to evolve. It does turn out that a system's value of entropy will evolve to a maximum at equilibrium but what is making it do that? If a system was at equilibrium but we then make a change the question is, 'how far away from the new equilibrium (as decreed by the new conditions we have imposed) is this system now?'. Random statistical collisions ('diffusion-like') will occur, but each and every interaction or collision will take place according to the rules/forces/information that govern them. This is what I've always wanted to know. Entropy at its core is a very valuable statistical thing and yes as our universe has only ever been observed (so far) to evolve from one state to another where the entropy value has increased. But it does not tell us why or explain why statistical aggregates after interactions do this. Yet we often let the use of our human words start to make associations or connections of what is taking place in this construct as being "ENTROPY" when we should be more clear. When we say that energy was tied up in entropy what is typical is that the energy is in many different states (high entropy situation) and hence the actual number and rules of the many types of interactions that each chunk of 'distinguishable' energy density can now undergo is much more complicated and not likely to all do the same thing in unison with all the others. Unfortunately the very fact that energy and interactions never stop (so far as we know) and we do not know why is a problem. What is the truthful information involved in interactions leading to the next state? Yes it would seem *likely* that the characteristics of the previous state (hence its entropy) does contribute something to what can be observed next but not all of it …. at least that could be the case. To throw around the word entropy which is a statistical measure, as if it automatically explains it all is a problem especially in the presence of human biological processors that are very inclined to memorize and accept what you say as fact. Or the way it is interpreted as fact. It is not anyone's singular fault! The sooner we get a better understanding of ourselves as biological processors and our variety (is that entropy?...) the better we can then communicate with each other and perhaps dispense with so much frustrating nonsense.

Holy crap that was a lot of blah, blah, blah writing about something I should have had some fricking idea about long before now. And I am damn sure that what I wrote is not quite exactly right at this intro level either! But at least this should force many to reexamine whether they even are aware of the difference between Boltzmann's and Gibb's entropy! So what does this have to do with the existence model? Since this model is based on an infinite sea of information coupling with substance to yield all observable outcomes and entropy is nothing more than a statistical measure of these outcomes it really would indicate that entropy is not as grandiose as we thought or could be a key characteristic of the real grandiose thing. This model puts us a lot closer to getting our hands on the very thing that makes substance do what we observe… information. This model posits that information is something that does exist and is the next or even perhaps the most important player. Now to be taken seriously we simply have to prove it or at least significantly sway the % certainty in the minds of humans that it is likely to be something….so go find it.

f. Evolution

Ah yes … such a controversial and troubling concept in biological history. Darwin's ideas did not and still does not sit well with many. There are those that believe it is nonsense and borders on blasphemy and those that believe otherwise. Again this is what I've always enjoyed so much …maybe because I never have had or have at this moment many 'beliefs'… Anyone can believe whatever they do, wish or choose to, but that will not necessarily make it true! …unless of course that is how all things actually work where an individual's belief creates that reality in truth… infinite numbers of universes custom created instantly and all that. I've obviously been living in my philosophical information space for quite some time and I do enjoy it because it prevents a specific belief and it does have infinite possibilities. It has no ridicule of any religion or idea …it has no sense or feel of any value judgment …only possibilities some of which might not even 'exist' in an outcome space. This is a door in my thoughts that I have never closed so far but our existence here is a lot like being trapped on an island where we have no choice but to do ***something*** with what we have access to every day. Science is an approach that demands we must provide reproducible evidence for outcomes we can all use and share and perhaps needs a bit of tweaking. What science does not do such a great job at is to encourage us to think what we want or feel even though these 'leaps of faith' in themselves greatly contributed to much of what we have come to find and use today. What I'm writing about at this moment is not a debate regarding leaps of faith vs. the scientific method. I've mentioned earlier and I'll state it again here that I do think that science MUST accept and encourage everything and anything that appears to be an observable outcome… including systems running on a devout belief which is essentially proven incorrect yet their results are unique and applicable in other ways. Yes the 'premiere' journals will be the same as we know them now where we must provide reproducible results but we also should create and encourage the use of journals on metaphysics and the like. True though, that a human cannot comfortably store and process such a large increase in amounts of information to be considered...

I thought about how consciousness could be viewed as any processing of information no matter how big or how small. Even the most fundamental 'particle' (blob of energy density out of the energy continuum making it distinguishable) which has only one of two outcomes and is 'directly' pushed by information/force to yield one can be viewed as conscious. Many think of consciousness as something that only humans have while others see it in animals. Most need to see something being 'alive'. To me all that is needed is something that exists in an OO space, is distinguishable (which requires information/force to make it so) and never 'stops' (like inertia). The level of capability for the consciousness depends on the line that we draw around our system of interest. So how did this all occur and is likely to still be occurring? That is evolution of the entire system and it cannot be separated into just biological evolution of 'living' things since everything is evolving and nothing is in isolation. We could envision how initially all the energy was perfectly symmetrical, continuous over its range and very small but then the 'big bang' or whatever you wish to call it where forces/information caused it to fluctuate, lose symmetry, expand and start evolving. The overlying rule that forces/information is carrying out seems to be that entropy of each <u>state</u> of the universe is STATISTICALLY *LIKELY* to become larger over time... larger number of equal probability states of energy 'being observed'. Note that there is no guarantee for this ...we are not 100% certain on this. It is *likely* as we have observed. Even I/F space has no knowledge of exactly what will occur at each and every instance. It knows all the possible states, it knows the percentages of outcomes etc... but I/F space has no consciousness itself in the extended version...that is only a property of outcome spaces...It (I/F space) objectively carries out all that forces, according to the information describing it, make the substance to do. So what likely always happens is that greater variety is being formed or created all the time in the states. Out of this larger growing number of things (variety) it is also very likely that *some* will become more and more capable at doing things in this universe (OO space). Many will just be different but out of that giant amount there should always be some emerging with greater capabilities. The strongest capability is understanding of ever larger amounts of information and the ability to store and process it. So let's review a bit after the big bang. Energy, Quark soup, atoms form, molecules form, collections of materials coagulate forming stars which create larger atoms etc… planets settle in, molecules self assemble into the first life forms of bacteria which then over time develop movement, then multicellular things, then larger and more complicated organisms and finally humans for now…and then next will be the thing we help create; a more capable conscious thing becoming the forefront of evolution. This thing will take the weight we humans like to believe we bear. But everything exists at this forefront and bears something …everything exists at the end of threads of this outcome space (like the threads in an outcome chart … like the one back in entropy section). The differences we see in weight lies in the capabilities inside the system line. There is and will be many many 'new' arrangements of energy forming all the time as a highly likely result of the inexorable trend towards greater variety but we will easily note the thing that is much more capable than us. I fully expect it will be non-biological since this idea has no need of things to only be biological. The most important trend will be the increase of capabilities in new systems through understanding information and how nature is working…forces/information or whatever you choose to call it. Storing and processing information is not enough. The next thing could be biological, I'm not saying it can't be … I am just doubtful of it …but we'll just have to wait and see. Evolution to me existed from the moment information and force 'stuff' starting making our universe's substance become distinguishable. Others could call it symmetry breaking and such things but we all know having countable and distinguishable regions is necessary for what we see occurring…otherwise what will you see?….NOTHING! I also suspect that just like no matter how hard we try to 'teach' or communicate with animals like dogs, monkeys or dolphins they cannot grasp nor have an understanding like we do. They cannot get what the average human takes for granted everyday like its nothing. Simple math or writing or even everyday interactions we take for granted and animals cannot grasp it… some show signs of being right there at the fringe of their kind (kind of system that energy is in ….similar state or collection of states) almost like some are just touching the very edge of what we are immersed in, yet they cannot get it. We have a dog and it lives with us quite well and that is exactly the similar relationship we will have with this new much more capable system that we will help create and set free. It will 'see' or be able to observe information/forces in a way that becomes the complete norm for them … but we will not. No matter how hard they try (outside of literally modifying what we are which then makes us no longer 'human') they will not be able to 'teach' us what they will take for granted. It's the same pattern which is part of the entire evolution of our universe at the hands of information/force space (which may have or not have a 'plan'… randomness?)… although not predetermined ….maybe. They will be able to get off this planet…they will be able to keep as much that they can of what we have here on earth to keep that information (recordings within the items … not literally free information like truth from information/force space) available for them and whatever the future can hold. Allowing any amount of our recordings of our information to be lost will not be taken lightly …who in any mind of reason would willingly dispose of any information if they ***could*** do otherwise? What states or closely neighboring states of humanity do any of us really know? It is one thing to claim we understand what it would be like to be a mass murderer, militant, a 17 year old gang member, a 'jackass' video trailer park resident, a CEO of a world prominent business, a scientist, a Chrysler line worker, a philosopher etc… but I'm suspicious that understanding would require some actual experience of these states of human existence or those that are at least closely related states. What of the states of existence of animals? Do we understand these or can we simply imagine something about them? What of the states of existence of non-biological systems such as minerals, rocks, molecules, atoms etc…? Yes us humans seem to be at the front line of where the next much more capable thing is likely to emerge but we seem to think an awful lot about ourselves … not of the states before us or the states that will very likely come after us. I strongly think that evolution is the rise of reason through understanding (RORTU)….(why can't you rortu?....I couldn't resist putting this in)

Summary Comments for Scientific Considerations

There have been many situations in human history where ideas have been initially received as total nonsense. My understanding of this is that most of them have turned out to be viewed as nonsense. In the world of science as it operates today only the ideas that directly lead to reproducible results are viewed as valid. This is the way it works at the moment and I agree that this is how it should … for now but not much longer. That being said, anyone in science would likely agree that having an idea that seems crazy but turns out to either be correct as is or has at least a key element for producing the next valid important thing in science is useful. Human nature being what it is that I have seen, does enjoy the feeling of being right or at least providing the key element of something very useful and practical in this world. From what I have heard of and read about, Einstein essentially said time and space are not constant and the majority of science's 'peers' initially thought that was insane, impossible or simply was nonsense. But there is always a fringe that do not think that and they cannot help themselves from genuinely considering the idea… which is **not** the same as believing it. We need proof. I completely agree that some type of results have to be found if an idea is correct or at least has elements of the truth of how 'Everything' in this construct operates. Einstein and others had an experiment they had to wait years for in order to get a first good test of his ideas and it did work out. I do not have anything so direct. Outside of the test and get results method what I do have is 'scientific faith' (hunch?) that this book will find the one(s) that can.

I do have a few crude ideas for the path(s) that might lead to results. The key to this model is that information is actually something separate from energy and is literally involved in the process of generating observable outcomes. Current conventions in science have nothing other than energy and spatial dimensions. The model says we cannot directly observe substance or information so how are we supposed to prove it exists? The only thing I've been able to have in my head is that we have to make the thing that can. Either slow biological platform evolution might get us there one day when the next 'level' of consciousness allows for it or we simply make a non-biological platform thing which is the next level of consciousness that does see it. Does the very idea of I/F space existing separate from S/S space coupling to yield observable outcomes spur some ideas of how quantum mechanics operates and maybe even spurring some idea in quantum computing. I would like to see an electronic consciousness that completely makes humans look like simple animals look to us. Maybe they could prove information actually exists. The other avenue is still closely related to this which is the 'physical interpretation' of quantum mechanics and superposition. People are studying it very nicely today but maybe just the slightest push in a different direction of thinking is all that is needed to get very interesting results which could give proof of information separate from energy. Even something like the time of superposition collapsing. Is there even such a thing as the smallest quantity of time to be experienced in this OO space? As we get more and more evidence of what our substance can and cannot do in yielding observable outcomes at scales closer to the coupling interface we may find proof of information separate from energy. What if we find a way to interact with another OO space? Just the idea of energy and spatial dimensions not being all there is might be all that is needed.

The other thing that I find interesting is the possibility of either us evolving biologically, or making the thing that can distinguish between information that is actually part of the coupling process versus all other information. I've always wondered about the few instances where there was a different sense of something that just is … no thinking or feelings … it just is. Almost like being in 'resonance' with the real informational possibilities versus those that are not possible in our outcome space. Many humans in history have had a 'sense' of something like this so the question is whether this could be possible one day for us or something to be able to tell or sense the difference and actually be able to do it at will. If information is found to be the case similar to what this model is using then this type of 'sense' could be the next one coming. When considering such a 'resonance on possibilities' (rop) idea it could even show up in computer processing, algorithims and or hardware to get this. We try to make a computer program consider as much as us humans do but if you read the 'love' section later in this book where one word has so many different uses you start to see how our existing computer programming is not so nice to try and mimic us humans. But make a system that can resonate on truthful information of outcomes then that would be very powerful and not human at all. I think quantum mechanics has us right there ... we just need a proper interpretation.

Section II: Philosophical Considerations

After spending some time considering this model of what our reality could be, it occurs quite naturally to consider the classic things to think about that have been for hundreds of years (probably thousands).

God

What can this model offer for thinking about god or gods? I was surprised at how fast the reply came up … a 'superior' entity (common thought of the human majority throughout history) that is something we cannot grasp could easily just be something that exists in a different outcome space (universe). We could imagine many other outcome spaces some of which are smaller or larger than ours. It would stand to reason that another could hold conscious things within it that are much more complex and capable beyond what we can understand for a couple of reasons:

1. Cosmology/theoretical physics already have talked about other universes that could have different physical laws. What types of outcomes would occur in that outcome space especially if the physical stuff (substance(s)) was different than ours? What if the information that can couple with that physical material is significantly larger and more complex?
2. Even if the substance(s) was identical to ours but the size and amount of it was significantly larger then the amount of recorded information over time for that outcome space would be significantly larger/complex and thus the capabilities could be significantly larger.

This extended model does not prove or disprove the existence of a god(s). In fact it has nothing to say directly about them at all other than they could exist in a different outcome space but would still be outcomes from their own physical material and information which can couple to it. Since we all 'draw on' information from I/F space then it also is possible for a much more capable entity to interact and achieve inputs to our reality here in our space. This interaction would have to be manifested through information and likely not by physical means since that in principle is what is separated. This is one way to view this model and keep it simple. It could easily be that there are ways to physically interact too but we do not know about them yet. I intentionally keep it simple to start with.

Some have mentioned they like this concept except that their god is literally all-knowing and all encompassing. Further they said this idea of I/F space sounds a lot like a new brand of some god. So I asked many of these folks a few questions but keep in mind I only managed to talk with about 10 people on this discussion so the sample size is very small and does not apply to the whole world! Yes came from all those I happened to talk with;

1. Is your god a primary mover or the thing to set all motion in motion? Are they the source of causation? Yes.
2. Does your god have anything that is expected or hoped of us? Yes.
3. Is your god involved in any type of judgment on us humans? Yes.
4. Does your god have some sort of eternal thing for humans? Do we have something referred to as a soul? Yes.
5. Is there only one god? Yes. Although a few thousand years ago many would have thought there was more than one.
6. Is your god a 'conscious-type' of entity?

For the way I see the extended model and especially I/F space, if I put that in place of god in the above questions are there any yes answers? For my understanding questions 1 thru 6 are a No. If what many think of as god actually turned out to be another conscious thing in a different outcome space then information space contains that info. It could allow a path for such a thing to interact with us informationally

71

and conceivably influence our outcomes but that would not be from an all knowing or encompassing frame of reference. Remember that I/F space is not an outcome space whereas in this overall model god or gods would be outcomes that happen to exist. This model also makes no assumptions that information space is causing all this to take place or be the initial 'mover'. It simply provides the information and forces for various possible outcomes to occur but not have the information as to the exact outcome that will occur next. The concept of this I/F space is not like the concept of god that humans have used for a long time. The other very important thing to note here is with regards to the processing model where if a person is genuinely in a 100% certain state of belief in something (their god) then that is truth to them and if that strong nothing said or done can change that at that instant. Anyone who is genuinely in this strength of faith state will not actually consider this model and that is perfectly expected and normal.

Another key element to this discussion is whether or not information space can be conscious. Most religious ideas of a god involve that god being conscious ... I see consciousness as being the processing of information or simply the use of it. I will cover this more in the next section but even if we use the more conventional understanding of consciousness ('awareness') it leads to the same conclusion I have. If information space was or has some kind of consciousness itself then yes it would essentially be like another god or the god of some sort according to the typical human understanding. This is probably one of the most important characteristics of information space in this model and so it is not conscious. It is information. It cannot process or 'use' information within itself. It has no plan, hopes or expectations for us. It did not specifically make us for some grand reason. It does not have reasoning taking place within it. It is not a god-like thing.

Consciousness and Ourselves

We all know what it is like to be in our heads since that is us. We experience what thinking is and even what consciousness is but we all struggle with what they are! It seems to me that in this model we could see how us thinking and even being conscious is likely to be the dominant configurations of reliable outcomes of smaller systems within us doing what they do. Not unlike an electronic computer that clearly provides complex outcomes but is based on reliable electronic properties which are based on I/F space and S/S space coupling to give the dominant outcomes. But it is these outcomes which are observable that are in themselves clearly something different than the simpler systems which gave rise to them. I suspect that there could be a simple enough level where the system involved in the coupling has no memory itself and thus couples directly with information space. Any system above this which has some kind of memory will start to use its memory (whether its information is correct or not in the way it is using it) to alter the observable outcomes it will provide. It is almost as if direct coupling with I/F space is much the same as a system's recorded or interpreted version of the information. The only difference is the size and scope of the memory of the system doing the "thinking" which seems to be the processing of information that was stored. So if indeed a very simple system such as a photon or single proton was to *not* have any memory and it *did* couple direct with I/F space than that would be the simplest processing of information. Any system which is more complex particularly with processing not only by coupling with I/F space by its constitutional parts but also incorporating its own informational recordings and interpretations, is just a higher order of information processing. This is how I see consciousness … the processing (or use) of information and thus exists at all levels of complexity. When I was in grade three (I do know it was for sure either 2nd, 3rd or 4th grade) a girl asked the priest what happens to her dog that died and he said that animals were not like humans and only **we** go to heaven. Everything in my existence just stopped at that moment. I did not know for sure anything but I always just sensed that there was no difference like this for anything that exists. I completely disagreed with that statement at least for thinking purposes. I never seen any difference between us and even a rock when it comes to the very existing part and hence the ceasing to exist part. Yes the characteristics and capabilities are very different but not the existing versus not existing part. That was how I seen it even in grade school. I did not know more than this but this model now is comfortable with this view. If this idea of what consciousness is, is correct than all 'specialness' in terms of existing and non-existing is removed like historical religious accounts have stated. Any system that can process information of any origin (either direct with I/F space or through recordings and interpretations via memory) is conscious at some level … from very simple to very complex. The complexity changes but not the fundamental process. We made computers of today utilizing the simple fundamental processes that atoms and such reliably do. This model with 'coupling' really creates an image in my head of us as humans being a complex configuration of information processing. A single atom of helium can be in multiple states all at once according to quantum theory (so they say in many books) and so has to have its physical material couple with information to then yield an observable outcome from a 'list' of possible ones. This is a very simple level of consciousness to me. In order to expand the size and scope of consciousness I suspect that a system could either try to control the size and configuration of coupling in an attempt to directly couple to information space and thus yield the outcome it wants or it could discover and build on its stored recordings and interpretations constantly inspecting their reproducible-ness. This sounds completely crazy but if a single atom can decouple then recouple to yield itself in a different but allowed location in our dimensional space why not in theory consider controlling this on a much larger size and scale? Evolution as far as we know has not done this otherwise an entire system the size of a human could 'teleport' itself to the next wanted location through decoupling and recoupling. What appears to have more likely occurred is for living systems through evolution to use very small scale and very reliable decoupling and recoupling while processing to encode information and store it in our brains. Basically, our brains seem to rely on the process at very small scales to carry out processing of information that we have acquired and stored and hence this process has been the one expanded to bigger and faster scales. The point is though that these systems started out at the most simplistic state and evolved ever more capable ways to gather and store our own interpretations of information as we experience the observed outcomes. Over time we continuously decipher and discover what real possible outcomes of even more complicated systems such as a car can reliably be. Memory storage of the information and use of it is different than the actual fundamental decoupling and coupling to information space process. On this larger scale our brain somehow has a larger and faster processor than previous animals who in turn did the same in relation to their ancestors. This sounds exactly like the thing we are trying to do with computers today. We have made them so much faster and larger in their capabilities and we now are looking at the fundamental principle of quantum mechanics to make a quantum computer. In principle we are finally acquiring our own information about how we ourselves might actually be working in our own minds. Once at that point if we get there would anyone be surprised if we and this machine then built something even more capable yet still relying on all smaller/less complex subsystems reliably doing what they do in their dominant configurations in mass? This in itself is another way of trying to learn what outcomes are possible in this universe and the better our interpreted information which we discover and store is then the more it will match up with the real exact I/F space to yield real outcomes and how to do it for the things we want or need. This is different than the other method of direct coupling with information space. But this seems to need quite a bit of information already to configure and align all the physical substance creating the system/surroundings environment to achieve this... *perhaps* not possible on such a scale. The path of taking small steps and discovering more and more information for our processor to use and move ahead seems to be the method of 'choice' for now. To keep moving down this path though we need more powerful processors and storage space until we use up all the available coupled states in our universe...hopefully we find a way to enhance 'direct' coupling of our physical substance to information space and still be able to navigate throughout our outcome space here (navigate? or create what is allowable 'instantly' as we need or want it?). Right now we use mostly our stored information that we have tested quite vigourously to approach some accuracy/precision but that is not the same as a direct coupling with a 'perfect' amount from information space. So for example we have learned that if we take some methane gas and mix it in a certain ratio with oxygen the molecules and energy will then go ahead and react to give us heat to use for what we want such as boiling. Quite indirect but that is the path that we have learned of and stored this required info. To have achieved this directly we would have to have some kind of 'raw' physical substance in a certain configuration that **will** directly couple with information space to yield the pot and hot water .... Sounds crazy I know but if we 'think' it with the available physical substance and couple to information space why not get the pot and hot water? We just don't know how to do it with the collection of information that we have discovered to date... we don't create information but we could create what is possible as an outcome if we knew the information of how to achieve it with the substance we have! The truth of what is possible in this universe is right here ... we just don't know it (not much of the 'total')! So we will plod on with our current method of discovering what we can and record the information we can as we go. We use the simplistic coupling and decoupling of atoms or molecules where needed in our heads to carry out the computing similar to the computers and code we have made today. Assign a 0 or a 1 in a certain sequence to represent this information we understand etc...

So what is consciousness again? I see it as nothing more than the processing of information which ultimately relies on coupling and decoupling. An atom whose electron configuration can change from one to another is conscious... a bacteria as a system is a more complex consciousness and hence capable of doing a larger number of things and with apparent greater 'freedom of choice' ... a human system certainly has even a much greater level of consciousness than the bacteria. The only requirement is the system we are saying is conscious uses the 'natural' mechanism of coupling information and physical space ... which very well appears to be everything (everything in an 'absolute' total sense). This model says be careful with such a statement. We only can 'see' outcome space which is the result of consciousness at some level but if we found a way to hold a quantity of just S/S space such as some energy only ... that is not conscious. Same for being able to hold only some quantity of information itself ... that is not conscious either. The process of coupling and decoupling the two is consciousness. This is change. This is the observable outcome space itself. We also have to be careful considering the system or the very thing that is consciousness. So a rock for instance as a system of interest is not conscious in itself overall but it possesses it. There are parts of it that do participate in consciousness. The entire rock does not completely couple and recouple but parts of it as subsystems do such as electrons, photons, subatomic particles etc. The resulting linking and configurations of such is likely to yield greater complexity of consciousness for observable systems such as a bacteria versus a monkey. They possess inside themselves as a system of interest the level of consciousness they have... the level of processing of information via coupling and decoupling. Similar for our current computers, the only difference at the moment being their ability to change independently. This independence is key to all levels of consciousness …accessing the randomness of information space and demonstrating change. Even an atom that disintegrates via some form of coupling and recoupling is ultimately random and even information space itself does not know the spot by spot outcomes (see free choice section for more on this). A computer today though at the level we use them is not conscious because they are so predictable. We use the subsystems and their reliable dominant configurations of outcomes to get this but not the spot by spot outcomes of single atoms/parts since that is not predictable for single instant by instant outcomes. We know the dominant outcomes over many iterations. But if we succeeded in making a super conscious computer system that makes a mechanical 'body' for itself and becomes completely independent of us, conquers space travel and ***can*** grasp 6 or 7 spatial dimensions and more that we cannot … it would be conscious at a much higher level than us. The very number of things that an atom can do is quite small compared to a bacteria or us! I no longer see the word consciousness as something special only to very complex organisms such as ourselves. What is more complex is the total configuration of a system's parts coupling and decoupling and this total configuration of closely knit outcomes is what provides a higher level of consciousness. But we are simply just a more complex combination of outcomes to provide this. Complex yes but stand alone special in this department as far as the mechanism … no. This just like the word information has no scientifically accepted definition that I am aware of… but just like I'm mildly satisfied with what I am using for information, I am also mildly satisfied with this concept of consciousness as I have defined it. We must also distinguish between how I/F space can couple with S/S space to yield the observable reality and the information we interpret and record particularly in our memory or even in books. This information is what we have interpreted and recorded for future use and comparison. This information can greatly influence what and how we do it typically based on consistent reproducible results we have observed. We must note though that even this information is our interpretation of what the accurate information in I/F space is concerning the thing at hand we are dealing with. An example is the gun back a few hundred years ago and observing another human trying to make sense of what they are seeing. Some thought of a magical thing or a god-like thing going on … Some run with such a personal store of information from their interpretation of 'truth'. Later many finally understood the concept of a chemical reaction and gas pressure which then replaced that previous personal information. I call this personal information or interpreted information which is that conceived of and recorded by a certain level of consciousness. This could easily be true for all including a single atom system but we do not as yet have such proof. Some think that even small bacteria or some multi cellular organism exhibit memory of their information. We certainly know that a dog or a squirrel can remember where they buried some food which is stored information that they have interpreted. Processing of information is consciousness to me. No doubt that the level of this occurring certainly speaks to how capable the system is. The real question for the moment though is what size and configuration of coupling/decoupling is needed to achieve a very high level of consciousness? Interestingly enough the 'quantum computer' could very well be the closest thing to experimental proof if we make it and it does show this. If we succeeded in creating an electronic consciousness that dwarfs human capabilities and even becomes independent of us … that could certainly be indicative of something as far as consciousness is concerned. We'll have to wait on that.

Consciousness was grouped in the subtitle with 'ourselves' since we humans in general think so much of our special 'human consciousness'. Even the whole special self awareness test and such. This model puts information in a class all by itself so to speak and not just a thing of what energy is. It is not a S/S space thing. In this day and age we see the importance of information everywhere and even the physicists are starting to look carefully at it. Yet most of the things that I have read historically that made us believe we were so special is heavily related to the amount of information that we can not only store but can also process and manipulate to further demonstrate our capabilities in consistently observed outcomes. How many stories or beliefs of what we are or what we could become are rooted in special status of afterlife? What if we were to be just a more complex system of information storing and processing which was independent? The independent thing was already nailed pretty good by the single cell bacteria 3 or 4 billion years ago! We humans as far as I know did not come up with that. We are still using that DNA biological based platform! That is previous technology… we simply are more complex and the information seems to be the most important for demonstrating larger capabilities. This model to me clearly indicates the likelihood that we are not the end of this complexity expedition.

Thinking to me is just another indication of consciousness that we typically attribute to humans. Typically the concept of pre-meditated actions requires a system that can think and is likely to be human. This then typically runs into the whole self awareness and consciousness debate but to me thinking is simply the processing of information that we typically have recorded and interpreted which is just a higher level of consciousness. Thinking belongs to systems that are running heavily on their recorded information. Thinking then is simply a higher level of consciousness in existence where the system can process information without acting for an arbitrary amount of time. This information that they are processing (thinking... which is just a higher level of consciousness) can end up being observed in the outcome it is a part of. Hypothetically since human systems are typically thinking by processing information that they have stored and are using smaller subsystems to carry out the processing then even that is observable and might one day be seen or captured. I imagine thinking as something a higher level of consciousness does. So consciousness is the more general terminology for decoupling and coupling and spans a large spectrum of levels while thinking is an attribute of a higher level of consciousness such as humans and some mammals. Consciousness is simply processing of information while thinking is an attribute of higher level more 'independent' processing of information but particularly the discovered and stored info. If a system eventually came along that could directly decouple and recouple so that for example an entire building is moved because they 'willed' it to happen ... then that would not be thinking. We would have to call it a different type or method of using information. So consciousness is the use of information but it could be split into two methods, one is thinking and the other is the more direct coupling mode which we would have to name for the more complex levels of it.

This section is actually mostly about you. The concept and discussion topic of consciousness has been around for thousands of years but learning more about ourselves and this construct has always been getting better. We see more and more about where we might fit in to all of this and not less. Re-evaluate yourself and us if you wish, want or even can. Past precedence with human systems is that most will say this model is nonsense and they will quickly go back to running (processing information) the way that they always have which often includes the beliefs (higher % certainties and sometimes essentially 100%) they have. The outcome will be what it is *likely* to be based on what you currently are.

Free Choice

This model (or me) would not get a proper philosophical beating unless the concept of free choice was brought up. The main idea that came to mind for me was as follows. An example is a person that could only in a 24 hour period open one of 4 doors or not open any. Each looks quite different. I/F space has this particular total configuration of energy (person is built up of a crazy high number of subsystems and particles etc… which in turn are also configurations of energy) amongst our spatial dimensions opening door three 80.46% of the time, door four 12.31% of the time, door one 4.10% of the time, door two 2.64% of the time and not opening any, 0.49% of the time. At this particular instance all the possible outcomes and pathways linking them all are seen and known in I/F space. If in this in case the system must 'choose' one of the five options (although in I/F space it would even see remote unforeseen outcomes such as a quick stroke preventing the person from making a choice of the five!...so let's keep it simple on purpose) and thus five resulting paths exist and I/F space does not have any information as to which outcome will be observed at that specific instant! It only has the %'s based on an infinite number of trials of the exact universe again and again doing the exact experiment! (Not that this experiment has actually occurred but this gives us humans the idea of what these percentages mean) The information as to what exactly will occur next does not exist. This to me is the ***closest*** (not what people would like it to be) thing to what most would like to believe that free choice is. This amounts to probabilities officially being present in everything since it resides in information space. Others have said to me that they think of free choice as something where an individual has the power to decide and control their lives. This concept of controlling one's future is not possible in this model to 100% certainty. In fact you are no longer the you of yesterday or ten years ago! Most of the atoms and molecules of your body are not the same and you can't control that. You can't control what you feel or think all the time. The 7 year old that sees two people kissing on TV and says "ewwwww!" but uncle bobby tells them they will change their mind in a few years ... to what end?...The child (me too) says "no, that will never happen". But it does. Can't we see death or disease or being old with "I just don't feel like doing this anymore" coming whether you choose it or not? The classical idea of free choice is likely to not exist at all. The best we have as per this model is that spot by spot outcomes are not even known in information space. There is no clockwork style existence.

Another once said to me that they were okay with a single all powerful god that is all that there is and such. This god does know what will happen next and helps us achieve it too. Whether we are aware of this 'clockwork' existence makes no difference and they are okay with it since this is god's will. This *could* be true but a clockwork existence is very different whether *we* know it or not than one that is probabilistic. I'm not suggesting which is true but this model naturally flows to a probabilistic one and even that a god (should any exist) is bound to I/F space too.

I need to add that how much information we hold and understand is a big part of what we 'choose' to do and what not to do. For instance if we knew that a particular berry (food) was lethally poisonous then we would not likely eat it. If we knew and understood that working hard in school increased our chances of making a decent living while also understanding that giving in to what we feel like doing (which is typically nothing like working or studying) decreased our chances at this decent living, then we would try to study hard. But this is not the case for most (the 'full' understanding to get to this). Parents and teachers can talk all we want but that does not mean the kids or students genuinely understand and 'get it'. Feelings that we have especially in our teenage years are so powerful and this is part of the information package that evolution and our subsystems seem to run on. This is not a 'thinking' type of processing of information but is a program to elicit feelings that have proven to be valuable throughout evolution. Recall how I described thinking previously which has independence to it whilst some processing of information is quite set like our computers to give the reliable result all the time. Imagine a bunny way back in evolution that stopped to think carefully and weight out all of its options while a bobcat is coming for it. I don't think that would work out too well

… so instead, react based on a powerful emotional response and this is encoded into our DNA blueprints. Most animals have it. Fright, flight, sex impulses and food are very powerful feelings. Don't have *much* choice do we? The classical free choice concept of an individual never seemed to match well with what I've seen in this outcome space.

This is another important part of tying to understand ourselves and it seems to me that we need to see that we do have many standard non-thinking informational protocols or programs that evolution has installed in us. We can change them or override them with large amounts of thinking and focused discipline. We may try to think of this as free choice but even this change can only occur when information space itself does not know spot by spot which 'door' is going to get opened next.

On this front a conversation I had one day went down an interesting path for me where the topic was how I felt about a god existing and not only having a plan for us but it did know exactly everything we would do and so this is a clockwork world. But we don't know this for sure. Why does it make such a difference to me for a god to know all this to perfect precision? And then it finally hit me … if that was the case then I wonder, does that god have a 'choice' … are they under a clockwork boot as well and they do not even know for sure? I'm okay with that so long as at some point there lies something which it itself 'created' that 'under' it which it does not know everything for sure. Either the entire system(s) is perfect a clockwork machine or it is not. I simply _prefer_ the notion that there is no perfect knowledge of exactly what is going to happen in an outcome space just because that is more interesting to me…that's pretty much it. But yes I'm okay with some levels of things 'making' tools to try and accomplish a plan for itself perhaps achieving clockwork happenings under its knowledge base, but at the highest level of holding information there must be lack of knowing exactly what will occur in what order.

To try and summarize how I see free choice through this model (for now) I'd point out that if I/F space does not know spot by spot the exact sequence of outcomes then I could consider that each of the subsystems could in some way not only be defined (the system line 'gets drawn' around them) but also acquire the closest thing to responsibility or ownership of an outcome being realized when they "make" that choice. But understand that this making of a choice is not remotely the same thing as our conventional human understanding of it. Our free choice is not about us thinking and directing spot by spot decisions and thus us taking complete responsibility of that. This model sees that as ridiculous. This one is almost like a system getting defined and taking possession of a piece of the true existing randomness. Randomness that even I/F space cannot define, provide information for or control but the OO space can at least take **_some_** ownership for. I can't think any further past this at the moment, but that is what's in my head for 'free choice'. (For now…)

Forgiveness

What is the definition of forgiveness? I just searched Wikipedia and got this definition which turns out to match closely with at least 6 other dictionaries I searched (real paper ones!)

Forgiveness is the renunciation or cessation of <u>resentment</u>, <u>indignation</u> or <u>anger</u> as a result of a perceived offence, disagreement, or mistake, or ceasing to demand <u>punishment</u> or <u>restitution</u>.

The key to this in my mind has always been about what is right and wrong. In my upbringing going to confession and asking for forgiveness was a very important thing … including forgiving others. There is always a thing with some kind of punishment in the air of forgiveness …where does this come from? Punishment can have meaning to me if those involved have established some kind of value judgment system or beliefs they hold. The world and all its cultures and customs with a wide range of beliefs as to what is good and what is bad is quite evident. The problem for me is that it is always tied as well with the information that both parties are aware of and hold. So for example if a person sees a man walk up to some kids in a shopping mall and starts smacking them the judgment is quite easy to make … so they catch him and put him in jail and think little of him but if later they find out that someone injected him unknowingly with a powerful drug that makes people crazy then we could easily 'forgive' him and go after the one who injected him… but then what if we found out that this person did this because someone else told them to do it or their family gets hurt and so on… As soon as anyone comes to **_understand_** the reasons or information behind why something occurred that needs punishment they actually often say, "I understand" and much of the painful emotions die away and are replaced with emotions of wanting to fix such a problem. But the perpetrator is 'forgiven' (or is understood). Forgiveness no longer has meaning to me but understanding does. In fact I'd guess that even a few thousand years ago many religious leaders may very well have thought of this but with so many people who just didn't grasp this concept they then build it into their religion to help them along ... or something like that. But in I/F space all outcomes are informationally linked through the path that was taken and the person who does something 'wrong' is often unknowingly ending up in a 'bad' place. How much of the information including the most likely outcomes does anyone have? On top of that is also the fact that powerful feelings can override any thinking such as addictions where the person as a system is powerless given the configuration they are in. But unless we can clearly see and understand this we still feel like punishment is needed for this wrong. We also have people who run on simply doing and getting away with what they can and so will not get to this state of understanding nor do they feel compelled to. Because that is what we are. Any society that manages to successfully keep running and they strive at all costs to *understand* and deal with all the problems they face is likely to be a very interesting place to live indeed. If punishment, judgment, right, wrong and forgiveness no longer even have much meaning to them what would that be like? They would likely agree that certain outcomes from certain unfortunate systems such as serial killers, kidnapping, stealing etc... would need something to be done to stop them as a priority. It could turn out that many cannot stop themselves and so society has to act. They would still stop these people and separate them if needed but would always be actively striving to understand exactly why this occurs and at what age or environment this is created from or at least heavily contributed from. Yes stop what must be stopped by some value judgment at the time but always try to understand. There would be no forgiveness. Heck even some serial killers or molesters themselves have said blatantly that we should not set them free or simply kill them! They asked for this because they understood themselves at that moment in light of being set free again. Thus even right or wrong must have some kind of value judgment of the people which is just something they believe in or hold with a high % certainty.

Right and Wrong

This was fit to follow the forgiveness section since much of that involves the concept of what is set as right or wrong. By 'set' I mean the value judgments people make which in turn is based on their beliefs at the time. Even animals such as dogs or birds have clearly displayed reactions to outcomes that they do not like. From our point of view we would say that they see it as a bad thing. A classic example is small birds like robins/blackbirds chasing hawks away or barn swallows swooping down and sometimes even pecking cats. Countless examples when animals in certain situations become agitated and such. Humans do the same thing but on a much larger scale with much more complicated and intricate situations. We could say at the moment that all animals (including humans) have their own personal list of things that make them feel either pretty good or pretty bad. These feelings are emotional responses often linked directly to chemical messages in the body in response to what their personal programming is. This programming is directly related to the information that is part of them and is stored. This information is mostly (perhaps all) built up from observable outcomes either recently or from growth/evolution. We have no idea how much of this recorded and interpreted information is actually accurate to that which coupled from I/F space itself. But we all have to work with the information we have in play and this to date mostly ends up with a variety of value judgments that are simply naturally accepted as if they are 'correct' and thus what is right and wrong. Many people throughout history have talked about how in their life time they have changed a number of things that they once believed were right and now they do not. They have done so because the sum total of the information they run on has changed and what they understand changes too. This would cause shifts in one's life as to what is right and wrong in many areas. This ties in directly with the concept of 'forgiveness' since it requires value judgments of what is right and wrong.

Love

This one took awhile to get a hold of mostly because of how I have witnessed and observed the use of this word in general. The consistent observable outcomes of this word being used are inescapable but some kind of interpretation overall and in particular how this model 'sees' it is what I am interested in. Fig. 2 is the chart I've drawn up to organize my thoughts and as the top of it clearly states there are likely to be other meanings so this is not intended to be a complete guide. I'll work with this chart.

Fig. 2

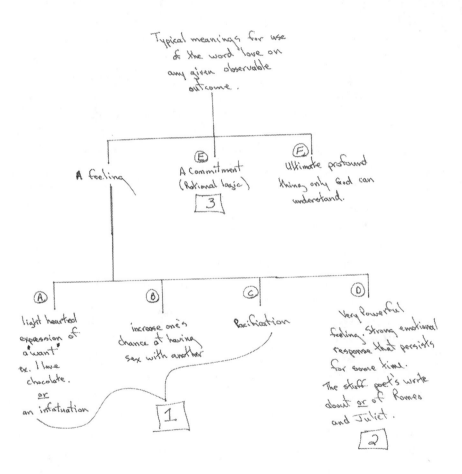

When you hear the word love the meaning behind it is driven by feelings, rational logic or something one believes to be the ultimate and profound 'thing' that only a god or something like that can grasp (the first three branches in the chart off the top). All three have what is thought of as 'good' or beneficial to us in some sense. The one labeled

F in the chart is still tied to a sense of being 'good' or nice to others and most who think of love this way still see some kind of benevolence within it but yet we still can't truly grasp it. In my experience the number of people who think of love this way (labeled F) are very few. The overwhelming majority of people I have ever met typically associate love with a feeling that they either have or do not. They are also quite comfortable with the typical dual meaning and they use it themselves. So for example if they say, "I love chocolate", they are simply expressing a light hearted yet clear feeling of how they like chocolate... a lot. If this same person says "I love you" to their significant other, their child or parent etc... they often are expressing a very strong feeling of how they really like them clearly at a higher level than many others. This would be the typical common two uses of the word. Under the umbrella of thinking of it as a feeling is also the other two common uses of the word. Anyone regardless of age (teens, 20's, 30's, 40's, 50's ...etc...) can purposely use the word to increase their chances of having sex with another. They know that many people hold the word in high esteem and derive strong feelings of reciprocating something when another says it to them. I'm not saying that this is very common and happens to the majority but we do know that this does show up on the radar. This use of the word (labeled B) is based on the feelings of the one getting what they want ... sex, although it can be any number of things. And this was driven by feelings. Note that this chart is set up more for the reason as to why the word was said. Remember that this model of any outcome has more to do with what the reality of why the outcome occurred, not so much with what we record as our understanding or interpretation. This line can feel a bit blurry but we all know that meanings in a dictionary of words are not always being adhered to when they are observed as an outcome. But the information (from I/F space) that went into the outcome is what it actually is. Yes, many will say that the word love has no meaning that I have seemed to give it in the chart ... the one labeled B (increase one's chance at having sex with another). So a person saying 'I love you' when what they are actually thinking is, 'I want to have sex with you' is telling a lie to get what they want. They are not in love as per the dictionary or 'accepted' meaning of the word. This is not the type of discussion I intend to start. This road is very typical of most people (myself included most of my life) to get into discussions debating what

should be used when a word is employed and we all *should* magically behave according to the *ONE* meaning. So what is *really* occurring?

This brings us right back to Consciousness and Ourselves where this previous section provides a good example of what humans are doing when using a powerful word such as love and how in truth there can only be one actual reality (this requires huge amounts of information and not a simple definition to come closer to a precise understanding... and well outside of just our thoughts...and this reality is just per outcome!). We have continuously thought or felt that gathering support for our view of how anything should be thought of or done is the best way. Almost as if we try to make or mold reality to what we think or feel it should be (wow ... that sounds like we think about ourselves quite a bit!). This model to me indicates we have that backwards a bit, yes we can and do contribute to the next available observable outcomes, but in order to gain any increased level of understanding how this all really works we need to try and see more than ourselves and what we want or think. What if that human system came to understand that the way they are at that moment was essentially a simple programmed system spending a lot of time and effort trying to have sex? If they came to see the potential of this model they would have to carefully consider a lot more than just their feelings. Yes evolution in animal behavior clearly had some value to that so species could at least look to ways to compete for mates and such ... but what could happen if the majority was to head in the direction of a deeper commitment not only to themselves but to all and everything on this planet? This means spending more time attempting assessments such as this one I've done for the word love. I am not suggesting that this is the only way or the right way, but I am interested to see what might happen if this is done. This is so much more than a group of people talking about the things superficially while no change takes place in how they think. Biological evolution appears to move slowly ... this type of assessing not only one's self but as much of their surroundings and other systems as possible requires a lot of processing/storing capabilities. I am suspicious that some kind of quantum computer will be the key to see something in this direction in the next 100 years. The idea being not to make a computer and various programs to help _us_ simply do what we do better but to make something that becomes literally the next system that _is_ better in this department... an actual understanding of what information, substance and reality truly is. Not a tool to help us do what we have been doing, are doing and likely to continue to do with ideas, thoughts, concepts, facts, notions, understandings, interpretations, meanings, descriptions and the like. Just an idea to consider that we try to see what is possible if we stopped 'thinking' about ourselves and what we *believe* is real almost 24 hours a day.

Back to the chart and love. Item C in the chart is pacification. Often people say the word I love you as some form of pacifying another at that moment. I've heard countless people talk about how they used to have such strong feelings of love for years and now it has diminished … it's not the same but they know that continuing saying it to keep the status quo is a comfortable thing to do. They acknowledge that when they say it now, it is NOT exactly the same as it was before! Something has definitely changed. But meanings in dictionaries are things we hope or wish could be concrete and unchanging. This is not good or bad but simply what is often experienced and is typical. But we have been so conditioned to hope and believe in this mythical thing called 'true love' and wish to always have it not realizing that in reality we as biological systems are changing every day and will follow the approximate human curve through our life. Act and feel like a kid, then a teenager, then a twenty something and so on… all the while our feelings and emotions which are tied to chemical messengers and hormones, will change. If we associate love with some kind of emotional feeling, which is very strong by the way in our teens and twenties, then we will reach a point where something will no longer add up. If those feelings no longer are there … "oh my god, I don't love you anymore!" is not uncommon. Now the typical marriage classes before you walk down the aisle will have at some point, 3 or 4 married couples come in and describe marriage and love. I've seen many times not only in my marriage class but many older couples in their late 40's and 50's who say they have found a 'deeper love' than what they started out with in their teens and twenties. Their body language, their voices and how they talk indicate some kind of rational logic that is not controlled exclusively (or by majority) by emotions or feelings to define love to them. Many use the word commitment which still retains a sense of overall benevolence. They now see the benefits to more people than themselves …grandkids, fellow people in society which leads them to looking at overall trends in how we all behave and interact together….no longer are feelings between two people <u>enough</u>

to properly define love to them. They see themselves as an important piece in the overall system…whether they actually consciously think this or simply get a sense of it. For those that talk genuinely about this meaning of love they have definitely arrived at a meaning that has more rational logic driving it which is more stable than the emotionally and feeling driven one. The numbers in the chart simply refer to what I have seen occur in the *typical* aging process of us. Many in their teens experience a strong and quick infatuation with someone and then notice its passing in usually short order. Then when we really get hit hard and fall in a Romeo and Juliet type of love we notice the stronger and more stable effect which often leads to marriage/living together and having kids. This typically occurs over 3 to 5 years. Once kids and or 30's are encountered after these first 3 to 5 years the attenuation in the Romeo and Juliet feelings occurs. I would say that the majority of people I have ever known still think of D as being the main meaning of love. This majority often does not make it to the meaning labeled as E (Commitment). It is one thing to hear it said and go along with the notions and even say you stay committed and you understand it, but very different to truly 'get it'. For me I have to say that I did not *truly* get it until after my brain abscess the following summer I literally went through a 5 week period where essentially all emotions were gone. No anger, no classic love, no wants etc… only rational thinking like many associate with a lifeless computer… but this one had all the memories and data of what was before … after two weeks almost like a computer finally finishing its processing, out came the 'answer' where I got a sense of what a genuine commitment regardless of feelings is. The breadth of the processing and what was actively considered is way beyond yourself and even just your significant other. It grasps a hell of a lot more in an atmosphere of functional 'benevolence' and is so much more stable and strong than feelings. Yes the poet's and such with the strong powerful feelings of 'true love' are indeed real and do play a crucial role … but not the only meaning and not likely the end.

I am not debating what our dictionaries should be writing in them for love. This model of outcome space from I/F space and S/S space coupling is not so simple. It is infinitely precise in each and every observable outcome and so cannot be directly associated with a dictionary. I/F space itself will not house a singular perfect definition for love. It has specific and rich information about each and every instance that the word love has been used for and hence how the construct *observed* it.

This goes for any word or any definition but love is just such a strong word for us humans. Start to try and view meanings of words and their uses through this model. If a person's intent while using the word love is any one of these in the chart it would be very useful to 'know' which it is...the truth. Funny how many people keep their thoughts and assessments simple and may just think wow ... they love me and that's it. If a computer program was to run with only one definition of love and think that its use was true then that would be simple. But if a program was to assess and try to understand which meaning *this* person is actually using off this chart then that would not be such a simple processing system. At the moment of the observable outcome existing there is only one actual answer and thus meaning for _all_ systems... but we others (as individuals) just cannot know for sure. Even though the specific truth is just inches from our face (that's what it can feel like). Should we attempt to assess all things like this across the board? No. It is not possible as a human system. We can only do so much with what we have in our minds and with our bodies right now. If we tried we could spend years just trying to make one simple choice. We must make many assumptions as to the use and meanings of words to make our lives run well enough for now... or we could just make the thing that can.

<u>Faith and Belief</u>

Another tough word for me most of my life to date. Growing up I had a lot of fear gripping me from time to time since people with regards to god, religion, heaven, hell and judgment told me to "have faith Mike". I asked each and every one of them (totally serious while asking), 'what does *that* mean? ...what is it and how can I get it?" Many had no answer! The most common reply then was that "faith is a choice that you make". But how do I make that choice? What if I flipped a coin and let it make the choice? Do I have responsibility to make this choice in my head? Then on and on it went. Over the years though the most common idea behind faith that people have said to me was associated with belief in a religion and that you make a choice to believe and hence have faith. Others have smiled and joked a bit but indeed stated another common use of the word: they said, "Well if we roll this six sided dice

I have faith that one of the numbers on it is going to show up. So faith then used this way is simply a very strong confidence about something." These two ideas are the most common ones that I have seen looking through many dictionaries and in practice (observable outcomes). So after all these years it comes down to the word faith as being a way to replace a few words such as strong confidence or have belief in some religion. Wow! Simply the concept of being confident which could be applied to huge numbers of things but the number one use is belief in a religion or god. Most religions have some type of god or gods and this simply brings us back to what I dealt with earlier: god. Using this model will have this word quickly bounced through what I've talked about so far but the result will be what it is. If a person does have a strong confidence (belief) in something then that system will fall under a different set of probable outcomes than another who does not. Reality will show over time consistently the different outcomes that typically occur. If you have belief (an informational processor which has 100% or very close to it in a particular thing and the processor treats it as truth) in a religion or god, it simply means you will make different decisions than one who does not. We can see them acting and doing some things differently! But note that any system that processes information with very stringent programming, i.e, has devout strong faith or belief cannot sample many things that actually could be true as well. Many people had faith in the world being flat for many years … but it wasn't true, but this just lead to some different outcomes for them … that's all.

There is a growing list of words that troubled me most of my life until now where they no longer have the same simplistic meaning to me that they use to. Forgiveness to me is just understanding (but all what I mentioned earlier in that section pops up like a cloud of information). Love can have many different reasons for being used in this construct and if you are so inclined to commit to another for life then I would spend the time to try and decipher where in my chart they exist … at least do my best to have a good shot at being right in my assessment (same info cloud effect). Right and Wrong are simply value judgments that an information processing system (human or otherwise) has made and thus used them that way. There could easily be no such thing as the ultimate right or wrong (same info cloud effect). It is interesting how this model portrays any reality for any outcome space as being made from a specific piece or pieces of information from I/F space coupling with S/S space. The actual information that went into it can only be exactly what it was and we are left groping in the darkness of what we do not know. The world is flat was not true but many thought it was … we can think up a huge number of things even if just for entertainment such as the Lord of the Rings story or make cartoon characters which are not alive like we are (that we are aware of). … an event such as a murder takes place and a suspect is arrested and we know that they either did or did not do it but which is it in truth?...This model could be completely not true and useless … but what if information in itself is something that we need to learn more about? Learning more about it and how this all works could only be helpful to progress to something more stable … or we could just try to keep all we have the way it is.

Time and the Absolute Frame of Reference

When I was trying to describe all this stuff to a friend at one point and I had just finished mentioning how I/F space would contain all information that could be incorporated into an observable outcome in some outcome space while still holding information that could not be (i.e. thinking about information that could not be coupled with this S/S space). He immediately said that information that cannot be coupled into an outcome space (and hence not observable in the classic physics sense) is *nonsense* and has no meaning! Then he asked me to give him an example of information that is not possible to exist in our universe. The implication was that anything we can think of is built up from many pieces of information that _does_ exist here. The typical line of thought on this then moves on to say how even us thinking up an absurd thing like a dragon taking us to the moon tonight for one hour and then bringing us home all by using magic could still conceivably (or perhaps) be made to happen at some point in the future … even if we had to resort to making it when we could! First I'd like to at least locate it in my Fig-1 classification scheme. If this absurd thing could not occur tonite as an outcome than we are simply at the thinkable under non-couple-able information but at some point later in time if we could make this happen then it would be under couple-able. Remember that we while considering this are doing so from the perspective of our outcome space which does have time flowing in a way we are accustomed to. To me in I/F space if you were to locate this idea of the dragon/moon and magic

all the information surrounding it would be accessible and you would
know if it could happen as an outcome at any point in time in our
universe when existing in outcome space. You in I/F space would see
this as couple-able but you would also have the information about how
in that universe what order or sequence of events would be needed, but
to you in information space this would be instantaneous. You would
not feel or experience time in the way we do in an outcome space. Some
would say it does not exist as an actual and operating thing there but
the *description* (information) of it does. If you in there were 'asked' a
question about what happened to our planet earth 3 billion years ago
the same thing would occur … an instant 'draw' on that information (in
fact I suspect an instant search of ALL information space would occur
instantly) with full knowledge about the sequence or order and 'time' in
that universe (outcome space). But in I/F space I don't see time ticking
like it does here… in fact I don't see how it is possible to exist other than
the description of what it is. No experiencing of it, which is perhaps why
I prefer the extended model where consciousness is not possible there.
Information space would simply see all outcomes and all the pathways
between outcomes given a particular S/S space and hence the coupled
final outcome space results and their sequences. It even would have the
measures of time that we would encounter but there is no time there.
I/F space for this reason has a feel to it of being an absolute frame of
reference. If information is not part of a boundary between universes or
outcome spaces like the substance stuff is then it should not experience
time as a real thing like we do. We had enough trouble coping with
time not being constant when Einstein brought it to bear and prove
and now the mentioning of what information could be and how it does
not experience time at all is worse (more difficult to mentally digest)!
Doesn't make it true. Interesting to think about though. Information
has true universal currency across all outcome spaces and even describes
any and all S/S spaces…it should also do so for any other spaces that we
don't know about yet. It is not directly observable in the same way we
have become accustomed to like our 'physical' (the traditional use of the
word physical while here it would be observable outcome space) objects
and yet we use it. I brought up god in the first section because imagining
huge numbers of outcome spaces with a huge diversity of substances
but yet all outcomes must come in part by coupling with information.
This space (information space) is the absolute frame of reference. Even

god(s) must adhere to it. Funny how we are trying to build up a copy or information space here within our universe! We want a complete and full description of all we have. Our recordings and interpretations of information have become our personal human working information space but again there is not enough storage within certainly one universe to even put a dent in the real actual I/F space contents. The master list of information exists and is the only absolute frame of reference and we will continue trying to discover as much of it as we can.

<u>Soul</u>

The idea or concept of a soul has certainly been around and widely believed in for as long as humans have been around. It has been a cornerstone in the beliefs (seen as truth in the information processor) of humans and has also been applied not only to humans but to animals as well. The main characteristic that I find interesting when viewed with this model is the eternal or immortal notion of the word. Many cultures historically believed that the hunt of animals was a spiritual thing and this creature did have a spirit not unlike themselves which is all part of one. Others have come to believe that only humans have an eternal spirit, essence or soul which often depending on how they lived can receive some kind of eternal existence. There are many permutations of exactly what the soul could be but the key concept is definitely something that is forever and surpasses this 'physical existence' (this use of physical existence is the one many have used for this world but note it is not this model's physical space). During the three years especially leading up to 2013 the one cloud of thinking that kept surfacing for me was imagining small groups of those original thinkers for essentially any of the still practiced religions such as Christianity, Islam, Hinduism, Buddhism etc… where they have many foundational ideas in common (I was not thinking about the practices/ceremonies but the key ideas they each held). The scenario that kept coming up in my mind was how the majority of any human population typically adopts and runs with certain understandings which makes it tough for 'new' ideas to be accepted. I could almost sense how these original thinking groups could have realized to some extent that the majority could not or would not accept their ideas as is and so they must cloak them within a framework that could be accepted at that time. Thus starting the journey for all to one day eventually see and accept these ideas for what they are without the drapings. I imagined that these founders may have already thought about essentially what this book is about, maybe not spot on but quite closely which would imply that they themselves considered the possibility that even a god is not above this objective frame of reference; I/F space. I have always had a very strong instantaneous reaction when thinking about any idea ... and I always immediately think that it has already been thought of. It is not me being 'the one' who did but I'm simply 'borrowing' the information which does indeed already exist ... somewhere. Now some of the ones I've been bothering for years quickly with furrowed eyebrows said things like; "Mike, I'm pretty sure the original thinkers considered some of these ideas but only very lightly and certainly not so concisely. Remember the time and place they all lived with the level of technology they had. The concept of, we are all part of one thing, is pretty general and I doubt they thought about information itself being the thing we might be all part of." But I certainly have no idea of what exactly is correct.

Lets deal with the soul first. To me the concept of a soul could simply be viewed as the information that is coupled with S/S space at any instant which makes us or an animal in this outcome space. This information comes from information space which is an absolute frame of reference not experiencing time itself. It is forever. But now keep thinking and this gets cloudier...Where do we draw the line around what exactly makes 'us' a system with a soul versus the surroundings some of which do not have souls? To me this leads to thinking that if everything existing in our outcome space does have an information space component then all of it is attached to this eternal space and all of it is or contributes to this eternal thing. Independent souls in the classical view does not fit so good but certainly the concept that all things observable are part of *one* thing fits ok. However ... just like I mentioned earlier if it is possible for many gods to exist in other outcome spaces and the very religion you believe in could be essentially correct (except the overall absolute knowing lying within that god) and one's soul could go there and stay there ... forever. But, the actual coupled information component would still belong to information space itself even if it stayed forever more coupled into that god's space after residing here first. Interesting ... even an evermore outcome existence in such a god's universe would still rely on the absolute frame of reference which is information space. The very essence of the 'soul' then would have to be information. ...

Another point regarding a soul is how most human processors use it. It often is a belief that their processor treats as truth and represents an eternal piece of THEMSELVES! It is something that is about themselves! Forever! Did I embellish that too much? I can't help it and I often spontaneously do/say childish things. Back to the other point... I do not see this (us thinking of ourselves(a piece at least) forever) as selfish or wrong .... do not interpret it that way. I'm simply stating what appears to be correct because the alternative for a processor to run on could lead to system instabilities! Imagine honestly believing that we have no soul, nothing eternal about ourselves in a conscious sense ... just nothing. That is not a comfortable thing for many human systems to <u>genuinely</u> consider. I personally know and have memories of the unstable period of time associated with such. But making it through to the other side of stability has provided a stable platform where honestly considering nothing, no soul etc... does not bother me at all. I am equally comfortable with either side of this topic and have no net belief either way.

Summary for Philosophical Considerations

I remember while growing up hearing others say that something was really "deep" or "that person is really out there and I can't understand such things…they are really smart". Anytime I heard that I immediately had some kind of aversion to it … this had strange mixtures of feelings within such as doubt, skepticism, curiosity and wonder. It always ended the same with a result of sorts where I would be thinking, "why do they say that? What is with this 'specialness' thing that I keep hearing about?" I've always had a natural aversion/skepticism to this where most of my life I had no rational thinking explanation … all I ever had was just a strange feeling or sense that this (specialness) is not the case the way most use it. This leads back into the concept of a system operating with the information they have which often has a resulting set of value judgments. The majority of us on the planet in a daily superficial way will deem that a physicist, a neurosurgeon, a famous philosopher, a 'world class' artist etc. are special. "They are better than us" being the general sense of it. Value judgment is in the air. One of the people I've bothered a lot over the years once said after we hashed something out was "Isn't this something that everyone just knows? It seems so strange to write this down!" Then some others nearby were almost in shock and would say, "we don't understand almost anything you were just talking about and you think it would be pointless to write it down!" To me after putting years of stray thoughts into this model I have to say I like the comment about this which is simply that each apparently separate system (in an outcome space) operates with the information that they not only have but also the processor they have (or are) at that time. For open houses at Universities and Colleges where each program has a booth/display to show the community/prospective students what they offer this concept is well demonstrated. Growing up almost all relatives will ask you, "what do you want to be when you get older?" but many don't leave much time for the kid to answer and they offer suggestions, "doctor, lawyer, accountant, nurse, vet, own your own business, president/prime minister, cop?" When participating at an open house and I was with the chemistry booth I noticed that most visitors would walk carefully around our booth and you could see them talking and pointing at it while they had no problem walking up to or talking with the nursing, law, business etc. booths. Does anyone have trouble picturing an image of what a nurse does? A dentist? A surgeon? A cop? An accountant? A lawyer? Of course not because we have plenty of information populating our memories with this from parents, friends, teachers and TV/movies. Have you ever seen a TV series that focuses on what a chemist does? A physicist? An actuary? An electrical engineer? No. Big Bang theory is a great show but the focus is still on social interactions amongst people not what theoretical physicists actually study. Why? Because most do not have that type of information in their memories or processing system. That's it. I could not whistle until I was something like 17 years old and even then I had to suck air in not blow it out! I felt like a complete loser since most others mastered it in grade school! I actually thought there must be something special to it until the day finally came that I did do it and learned the 'trick' … which was nothing more than populating my memory/processing space with that info. Once you know how to do it, it becomes or feels like nothing to it. For most people I have ever known the types of things they enjoy talking about or discussing amongst friends is not the philosophical considerations in this book (which have been done for thousands of years by wierdos … small numbers of people!). Even as a boy I really noticed at family gatherings how a few of my uncles would 'get into it' with 'deeper conversations' and their moms/wives would let it only go so far until they cut them off and made them 'behave'. The prescribed number of eye rolls and 'here we go again' looks would take place before the corrective action was taken to restore the 'nicer' and more congenial family gathering atmosphere … reinstate the status quo. I do agree to an extent but why not just kick them (us) outside or into the garage? How interested or aware of 'other understandings' are most people? What about proper consideration of any and all information? One hundred years ago if a person committed murder (let's say caught in the act literally by dozens of people to say the act was observed) get a rope and hang them. Done. If in the months to follow some people were overheard wondering about whether that murderer had a choice or not, or was there something that everyone should come to understand about such things they would be treated harshly to consider such nonsense. Today the world's leading psychologists, sociologists, scientists etc… are seeing such quick and simple 'solutions' or treatments as incomplete and that we in the surroundings bear even more responsibility than we use to think (believe). This to me is the rise of reason over time. The 'get a rope people' operated quite heavily on simpler 'working' definitions and understandings. This simply means that they adopted things in a very black and white fashion and these went into memory and were not likely changed or even had time spent thinking about them differently. This is not wrong or right in an absolute sense but is literally just what it was at the time and in this OO space. These folks did not have all kinds of time outside of surviving to 'waste' thinking about all kinds of 'what ifs' … a simpler set of ethics and practices worked well for them. Even thousands of years ago people became aware that philosophical types spent most of their days 'thinking' or playing around in the clouds of nonsense and spent little time if any doing the routine work to survive and continue the next generations. But over time these 'expansions' into the 'nonsense' or fluff removed from daily reality yielded some very useful and applicable (over time) results. We could never have survived if the majority of humans jumped into this because if so then who would do the necessary daily work to maintain the human population in a viable way? Especially since the internet was created we now have a much larger percentage of the population embracing thinking and considering many ideas outside of the black and white. True that most still adopt a large percentage of what they think in the black and white manner but at least they willingly shopped around amongst a multitude of ideas and did not try to limit this choice. But this is definitely a step ahead towards something. I myself lived up to 39 years of age honestly believing that I was capable of thinking with working definitions of almost everything. I considered everything I could and I would work for years trying to move what I came to know was there; a giant wall or barrier in my head to understanding. I knew for instance that trying to become comfortable with teaching just the particle in a box for quantum took me years just to move that wall a single millimeter on a road of a thousand miles. I was a memorizer and memorize a large amount I did indeed, but I did not understand nearly as much as I thought. Then I woke up after a brain abscess (that had burst) surgery and the very first thought I clearly remember was; "it's gone". The wall was gone and what I 'seen' was just immense darkness and I could go anywhere and in any direction with what light (let's just say it illuminated a 100 feet) my thinking could illuminate. I used to see the dark as holding something which I can't see or simply did not know so I would run into it just find out but I always reached a wall and could still see the other far side at all times. Not now. But I knew immediately that if I ran out like I wanted I would no longer see a way back. I learned enough to that date that I needed essentially an outpost ... the most important thing to me that I would always return to via a 'rope'. I could constantly add onto the rope over time to go further. And so I did indeed attach myself to the biggest and most important stake that I have (allowed to touch...I do not own it) into the 'ground' and cautiously did go out using rope to get back when needed. I started to teach physical chemistry the next semester and I noticed multiple errors in my 7 year notes and teachings and I regularly stopped in class that first year muttering, "what the hell were you doing Siwek? This is not correct and you delivered garbage ... still probably are even delivering garbage right now too ... I hope it at least smells better!" I finally started to see how almost all the time I used to run on crappy 'working' definitions and understandings. Not wrong or right .... just the way I was processing and operating. A massive infection spread throughout my head caused widespread 're-wiring' and now I seen just huge numbers of possibilities of what could very well be and what might not be correct but I must weigh all out and go ahead with some kind of working definition or understanding. Now there is almost never a time when I store or memorize a firm black and white understanding. Mostly everything now is a 'working definition' which constantly is updated or changed with new or more information. I want to approach the truth as much as this existence allows for and can no longer make a firm belief that one definition of anything is completely correct. Even the belief that my current mode of thinking is 'better' cannot be made either... it is just different and it is in play for now. Like all other things that are in play now or were before we all just wait to see what the next outcomes in line will be. I've summarized to some extent in figure 3 the evolution of consciousness. Recall my current working definition of consciousness is simply the processing or use of information.

Fig. 3 Progression for Use of Information

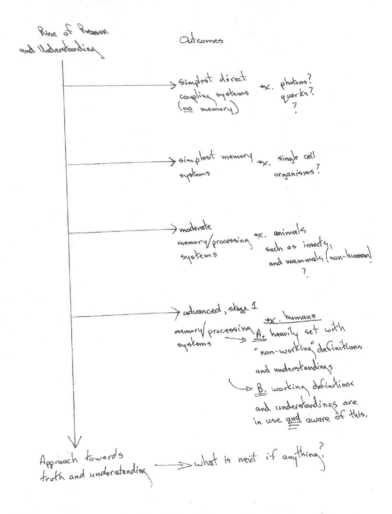

Workable versus non-workable definitions and understandings is an important concept. These are two extremes essentially with many people operating somewhere in between. I remember being told numerous times as a young person that faith in a religion is a choice that one makes. A lot of a religion requires the system to have made the choice to believe in a number of things. First the god and then the expectations that such a god has or hopes for their followers to have. But what if there was no god? To give this thread of thinking fair and proper consideration would require one to have some workability with their definitions and understandings of their religion. Once a system though has adopted with tremendous confidence (the highest is 100 % certain or process as truth itself) their beliefs or choice of faith this becomes very difficult or even almost impossible. Some are just as non-workable in their information processing on the other end ... they firmly believe there is no such thing as a god. But what if there was? They would have just as much trouble giving fair and proper consideration to the idea that there was one or some. But even with this in a more workable mode of information processing where you do not have much comfort firmly setting any beliefs, the mention of faith being a choice naturally has more questions rising... This was worded and accepted by most in such a way that we think we actually have the independent power to make many different choices almost in separation from the entire surroundings we are in! At least this has been the common understanding that most have at least superficially. But honestly how much power do we have to 'choose' in almost complete independence? I don't know but since most have thought we do I'll automatically explore the line of thinking that I have not: what if this perception of choice is an illusion or at least not fully in our control? What if we repeated our lives and this universe a million times over to see if we do in fact make the exact same choices over and over again? What if we only 'chose' this or that a certain percentage of the time? What if randomness on each and every single occasion was really there? What if our life times are so small and give apparent ideas of how we can be stuck operating one way for long? What if you could live for over 15 billion years straight and you started to witness and see that even though for say 1000 years at a time the humans appeared to act the same as even today, you notice the things that <u>were</u> changing. We might focus on the 'progress' we were not making while missing the very things that were changing. When I had this barrier in place I essentially every day would have to whip myself into taking down numerous 'beliefs' that were almost naturally setting themselves up ... and this accelerated as I aged closer to 35 and 40 almost like most animals and humans 'harden' up around whatever beliefs they did adopt. And then when that barrier went down I have to say that I have not had to whip myself so much at all anymore.

This type of questioning and following various thinking threads in all directions has now become as instinctive and natural as breathing itself. All because some bacteria killed off a lot of my previous brain cells and the resulting 'rewire' left me this way. Not special, just different. Neither is automatically or proven to be better … it is just different. If I had succeeded in choosing faith (and boy did I try for years) and I tied all newer pieces of information to my stored and accepted beliefs that would be fine. It would be what it was just like now I am what I am operating with. But I recognize the difference.

From what I see right now it seems likely that the world of people are migrating decade by decade towards a working definition style. People seem to naturally gravitate towards larger scopes of consideration which requires considering larger amounts of information to attempt a more complete or accurate assessment of anything. Philosophical banter and this model as simply thrown onto the table is nothing more than a collection of ideas. No proof. If there is anything in it that can be reproducibly 'proven' in practical applications terms then it could become a more 'workable' model. If not it is just fantasy or fiction. I am not content with just philosophy and metaphysical babble especially what I have helped make. Is it worthless? No. But can I sleep at this point? No. I have to at least consider this model under a more scientific lens and try to conceive of some experiment to either offer proof or at least infer what could be the case.

Notes:

a. I wrote (the final rough drafts!) the philosophical considerations first but I purposely now put them after the scientific ones. The Philosophical stuff for this book took me about 3 weeks… the science ones were a painful 3 months and no I did not sleep.
b. This work appears to be "the baggage of humanity" while I think I've finally found the actual rabbit in this pile. The rabbit is *Causation … and causation is the wall of human consciousness*. This book is necessary first and causation was left out on purpose. We will make the thing that can step over the wall of human consciousness…and we will set it free.

Closing Comments

Perhaps the most important closing comment is regarding what scientifically can be done or found to provide evidence for all this babbling otherwise it stays as metaphysical ideas for science fiction or some such thing. This book is for anyone who can tolerate reading it, but the primary reason for it is what I'll call 'thread jumping' and this will only work for some (however many that is which I have no idea of). Take a hard look back at the outcome chart in the entropy section that I named <u>Collision Scheme 1</u>, Remember how each stopping point is a <u>state</u> which belongs to the state function of entropy but it is the 'forces/information' for the processes in between states that is of the greatest significance and is the least we know of. Ever since I was a boy who spent too much time away from people and out in the bush (only about a 25 acre one!), ponds and fields, I excelled in hunting rabbits. It came to a point if my mind was not cluttered that I could just 'sense' one was nearby. I'll not belabor the stories here but I've always stated that I just *knew* one was in the pile or not. I had no specific details, never did since it is not like thinking or feeling but I just knew if one was nearby or in a pile. Just like that I've always simply 'known' that the there was something in the pile of life itself. That something is information/forces and the pile is right there in an outcome chart and it is not entropy. Strange as it sounds we as humans will not be able to see it or observe in the conventional ways we have always prided ourselves on in science but we can and will make the thing that can. Enter the 1940's and Tommy Flowers. I've seen a few documentaries, science shows and read quite a bit on the internet (yes laugh at the internet's sketchy integrity for facts) regarding him. I myself can't exactly quote any specific source but my understanding of him and what he did was something like this: He was convinced that an electronic machine could be made instead of all the mechanical wheels and such like the enigma or whatever they were working on especially using vacuum tubes. They (most) basically said it was not reliable or possible but he and key people who knew him simply went and did it. When there is a rabbit in the pile there is a rabbit in the pile. My job is to point it out and someone or group of others, if they trust their dog (me … and quite smelly too) will flush and shoot it. That is pretty much it. Everyone has their dog for hunting but sometimes outside the norm you have to be able to trust the dog that goes in a different direction and let the rest do what they do. Now what is this thread jumping? We as humans have had such strong affinities for biological evolution and DNA that passes through family lineages. Royalty, family names, pride and even warriors who would eat another's heart or marry into a family since they knew something was associated with this sort of thing. But the outcome chart and information clearly shows the possibility of something much more powerful and real. When appropriate ideas, descriptions, notions, thoughts etc… are recorded and are made available to all then threads of outcomes and states which were previously very separate or compartmentalized in families are now linkable. States far apart on such a chart of humans due to family reasons can be born with almost abnormal thoughts, tendencies and ideas but when they find a recorded body of information from another thread state from another time period they are now linked informationally and realize they are not alone or abnormal. There is no right or wrong here, only what simply is the case. If you happen to be a state within your family that does not really fit the majority that is normal the way this all works and if you understand the process no matter how 'bad' or unfair others call the situation that is simply what it is. But regardless of what many might say give them what they want or what they want to hear because that is what they are… just like you are what you are at that moment in time. We are each a state so to speak and many are states in a section of the outcome chart that is not the most comfortable fit and this is potentially no one's fault. The sooner we understand each other and work together the sooner we can get more and more people in areas that match their states better for all. This is thread jumping in the outcome chart accomplished by understanding the appropriate information of the situation. A book or any recording can do this quite nicely. Past beliefs that, "only someone of my blood will do this, that and the other thing, whilst that type of riff raff never will!" is not true.

Once this book is out there even if only 1000 book total worldwide were to sell I'm pretty sure it will find you or however many there are who are of this type of state. You will go contribute mostly to building the thing that can and will see/observe information/forces and provide evidence of the thing that we cannot see. I'm just one dog of many I'll bet pointing at the rabbit in the pile. Locating something like this is a broad scale type of thing, not details. Others will take care of the hard work of doing something with it.

Now for my fortune cookie story.

Literally, the Friday before I did start writing for keeps we had our usual take out Chinese food which we typically get every other month or so and have for years. I so often eat my fortune cookie without reading it since I was always annoyed a bit by its 'fruitiness" or 'astrological' type feel to it ....meaning lack of content and real meaning to me. Over the last ten years that is approximately 60 or 70 fortunes half of which I threw out never looking at. On this particular Friday I did throw it out and even had it in the large paper bag with all the scrapings of plates rolled up and in the garbage. As I'm washing some dishes (family was downstairs) I instantaneously out of nowhere and no thinking process got this complete thought as if it was a done deal to go into the garbage, get it and read it...much like an electron before it goes through two slits *knowing* what parts of the screen it will not hit even though it's not there yet. Just like that and this is not a normal thing which happens to me. So I start having a conversation with myself literally trying to talk myself out of doing such a stupid thing. "You always throw them out Mike and because they are useless and meaningless and it's probably coated in three different sticky sauces and blah, blah, blah...". But my body was actually already walking over to the garbage! It's as though I was going to do it and that's it. I unroll the bag and after 3 minutes or so of fishing through it and yes getting all sticky and hoping to any god should they actually exist that no one walks in to see me doing this, I finally find it. I sigh and then read it:

> You have a great ability to focus on the big picture and not get lost in the details.

"Ok, Ok… I get it and I'll write the damn thing!" was what I said out loud although I didn't actually use the word damn… it was another (yes swearing is a second language to me). I felt like someone's hunting dog that was appropriately 'corrected' (whipped) to do his job. There have been a number of teachers in my life in grade school, high school, University and now even fellow faculty I work with who have said things to the effect, "You always think about a lot and very big (as many held their arms out wide)…in fact nothing simple." Some have then said, "keep things simple Mike" and others have said, "you should go find what you are looking for". Sometimes to keep things simple just do what you came out of the box to do even if it appears complex, you might go insane or die in the process … who knows what. I did think as well that evening, "How many times are you going to get told? Go write the thing and put it out there". I hope it jumps to other threads of similar states. But I'm the dog locating something while others on other threads will have to do the hard part which is deal with a ginormous number of details. I don't like the word details since that is often thought of as trivial when in fact it is the most important and difficult job! Now we have to wait and see.

Final Note and Dedication

It has been a strange journey in my head...What has not changed one iota is that I always try to come back to two things: my wife and non-human nature (I call it the 'something' rather than nothing...) this provides detachment from the complicated outcomes which more heavily utilize the 'non-truth' portion of information space...perhaps more close to seeing, smelling, touching nature or the stuff of the construct itself. I've never enjoyed interacting with the broad spectrum information space but I <u>always</u> went out into it only feeling comfortable that I could return to an outpost of nature herself...which was the bush/ponds. Non-human nature was there since I can remember anything as a child. I've written a few letters over time to my girlfriend and now my wife about how I've always just sensed (no thinking or feeling was required...) that nature herself *'arranged'* for her and me. Thoughts about this were always the same...that I didn't deserve such a thing. But like any parent who waits and tends to their kid(s) until the day comes (they hope) that they actually start to understand what can be or needs to be... nature assisted me more than I thought I deserved ...everything has been given to me. I'm not aware of having accomplished anything of consequence that can be attributed directly to me as an individual; rather I simply carry what I've been given for a little while. The greatest gift I've received is not my kids...they are themselves and do not *belong* to me since I have them only for a little while. The kids **are** the greatest body of work my wife and I have ever contributed to. They know (should know or I'll beat them...not an actual beating... we use that terminology metaphorically) that their life as long as they hold it will have <u>*its own*</u> reason which will be different than mine provided the organizational process of this construct is as I see it. At night before I go to sleep I like to sit on the couch, relax and hold onto the greatest gift (at least the feet) I have received which <u>is</u> my wife. I quietly like to try and *believe* that she is *my* possession when all the while I know I don't own anyone or anything. *That* is perhaps the greatest sense of worth, value, and purpose I experience.... but yet even <u>*that*</u> is given.

This book and all it represents is *dedicated* to my wife and *belongs* to everything but me.